Beginner's Guide
to The Fungi

Beginner's Guide
to The Fungi

C. L. Duddington

PELHAM BOOKS

First published in Great Britain by
PELHAM BOOKS LTD
52 Bedford Square
London, W.C.1
1972

7207 0448 0

9001566570

Set and printed in Great Britain by
Tonbridge Printers Ltd, Peach Hall Works, Tonbridge, Kent
in Baskerville eleven on twelve point on paper supplied by
P. F. Bingham Ltd, and bound by James Burn
at Esher, Surrey

Contents

CHAPTER

1. Introduction   11
2. Mushrooms and toadstools   24
3. Bracket, tooth, club, skin, and jelly fungi   43
4. The toadstool and the tree   54
5. Puff-balls, stinkhorns, and bird's nest fungi   62
6. The rusts and smuts   69
7. The cup fungi   82
8. The flask fungi   93
9. Powdery mildews and green moulds   102
10. The yeasts   115
11. The grey and white moulds   128
12. Fungi that hunt   140
13. The lower fungi   152
14. Collecting fungi   162
    Glossary   167
    Selected Book-list   172
    Index   173

Acknowledgments

Plates 1, 7, and 15 appeared in some articles of mine published in *Teachers World,* and I should like to thank the Editor of that journal for permission to reproduce them.

List of Illustrations

PLATES   *facing*

1. Photomicrograph of a section of the gills of a mushroom (*Agaricus campestris*).   32
2. Photomicrograph of a transverse section through the pores of *Polyporus*, showing the hymenium lining the tubes.   32
3. Photomicrograph of a transverse section through three of the spore-bearing spines on the cap of *Hydnum*.   33
4. Photomicrograph of crushed roots of heather (*Erica* sp.), showing the associated fungus.   33
5. Photomicrograph of part of the cortex of an orchid root, showing the mycorrhizal fungus in the cortical cells.   64
6. The common puff-ball (*Lycoperdon perlatum*).   64
7. Photomicrograph of part of a section of the fruit body of a puff-ball.   65
8. Photomicrograph of a perithecium of *Sordaria fimicola*.   65
9. Photomicrograph of a vertical section through part of the tip of a mature stroma of the stag's horn fungus (*Xylosphaera hypoxylon*).   96
10. Photomicrograph of a vertical section of the stroma of the ergot fungus (*Claviceps purpurea*), showing the embedded perithecia.   96
11. Photomicrograph of the bakers' yeast (*Saccharomyces cerevisiae*).   97
12. Photomicrograph of a zygospore of *Absidia glauca*.   97
13. Photomicrograph of an eelworm captured by the sticky networks of *Arthrobotrys robusta*.   128
14. Photomicrograph of two eelworms captured by the constricting rings of *Dactylaria gracilis*.   128
15. Photomicrograph of part of a transverse section of a stem attacked by *Peronospora*.   129

TEXT-FIGURES
1. Diagram showing the types of flagellation in different kinds of zoöspores.   17
2. Part of a conidiophore of *Verticillium* bearing whorls of phialides with conidia.   19
3. Five stages in the development of a basidium and its basidiospores.   27

7

8 LIST OF ILLUSTRATIONS

4. Diagram of the formation of the ring and volva in the fly agaric (*Amanita muscaria*). 31
5. Diagram showing the different forms of gills found in agarics. 33
6. Four toadstools of widely differing shapes. 35
7. Block of wood attacked by *Merulius lachrymans*, showing splitting along and across the grain. 45
8. A fruit body of *Hydnum*, showing the spines beneath the cap. 49
9. Types of basidia found in the jelly fungi. 52
10. Fruit body of the common stinkhorn (*Phallus impudicus*). 65
11. Diagram showing three stages in the discharge of the spore howitzer of *Sphaerobolus*. 68
12. Stages in the life cycle of *Puccinia graminis*. 71
13. Teleutospores of the Ustilaginales. 80
14. Diagram showing how the ascogenous hyphae and asci are formed in *Pyronema omphalodes*. 85
15. Some cup fungi. 89
16. A stroma of the stag's horn fungus (*Xylosphaera hypoxylon*). 97
17. The ergot fungus (*Claviceps purpurea*). 100
18. Powdery mildew of the rose (Sphaerotheca pannosa). 104
19. Tip of conidiophore and conidia of *Penicillium*. 106
20. A conidial head of *Aspergillus*, showing phialides and spores. 112
21. Vegetative reproduction in yeast. 117
22. A sporangium of *Mucor mucedo*. 130
23. Sexual reproduction in the Mucorales. 132
24. Sporangiophore and sporangium of *Pilobolus*. 135
25. *Stylopage cymosa*, showing captured amoebae and conidia. 142
26. An eelworm captured by *Stylopage grandis*. 143
27. A dead amoeba containing a thallus of *Cochlonema verrucosum*. 145
28. A sporangiophore of *Plasmopara viticola* emerging from a stoma on the surface of a leaf of a grape vine. 155
29. Zoösporangia and zoöspores in the Saprolegniales. 159

Preface

In recent years there has been an awakening of popular interest in the fungi. More and more people are taking notice of the toadstools that grow by the wayside and wondering where they fit into the plant kingdom. Mycophagy – the eating of toadstools – also seems to be on the increase. These things are all to the good, for the fungi are a most interesting and rewarding subject of study.

The fungi comprise much more, however, than the mushrooms and toadstools. The vast majority of the fungi are microscopic and tend not to be noticed unless they intrude on us, either by mildewing our rose bushes as does *Sphaerotheca pannosa* or, on a more beneficient plane, by supplying us with something we want such as the drug penicillin, which comes from *Penicillium chrysogenum*, or beer, a product of the brewers' yeast (*Saccharomyces cerevisiae*). Yet these small fungi are just as interesting as the larger toadstools, and in many cases even more remarkable. Few people have heard of the predacious fungi that trap eelworms alive, or of the strange 'spore howitzer' of *Sphaerobolus*. It is in the hope of interesting a few newcomers to mycology in something more than toadstools that this little book has been written. It does not pretend to be a complete account of all the fungi, for that would need many volumes each of much greater dimensions than this one. I hope, however, that, by selecting a few examples out of each group of the fungi, I may encourage some of my readers to extend their area of interest to include the smaller fungi as well as the mushrooms and toadstools. They will find it well worth while.

C. L. D.

Fovant, Wiltshire

1. Introduction

The study of fungi is called mycology, a word that comes to us from the ancient Greeks. According to the Greek legend, when Perseus returned to Argos after securing the kingdom by murdering his grandfather Acrisius, he felt so ashamed of his deed that he persuaded Megapenthes, the son of Proteus, to exchange kingdoms with him. This done, one of his first acts was to set up the city of Mycenae. There are two stories of how this name arose. One is that Perseus lost the cap *(μύκης)* of his scabbard at the place where the city was built, while the other is that Perseus, feeling thirsty, picked a mushroom, also called *(μύκης)*, and drank the water that flowed from it, afterwards calling the city Mycenae in gratitude. Whether the 'mushroom' from which Perseus drank was the same species (*Agaricus campestris*) that we commonly eat today, or some other species of toadstool, legend does not tell us.

Out of the nearly two million different species of organisms that inhabit the earth, between fifty and a hundred thousand are fungi of one kind or another; exactly how many species of fungi there are it is impossible to say, for one man cannot know them all. Some, like the mushroom, are familiar to everybody. The effects of some of the fungi are also well known. Most people have seen apples destroyed by the brown rot fungus (*Monilinia fructigena*), and may themselves have benefited from penicillin, a product of a fungus called *Penicillium chrysogenum*. Such notoriety belongs to only a handful of the fungi, most of which are microscopic species that, although they may be common enough, lurk unseen and usually unsuspected, and are known only to mycologists.

What are the fungi? A botanist might define them, rather loosely, as 'Thallophyta without chlorophyll', a definition which may call for a little explanation. The group Thallophyta includes those plants whose bodies are not divided into stem, root, and leaf, and which have certain other characteristics which are purely technical: in common speech they are seaweeds and their fresh-water allies, the fungi, and the curious compound plants called lichens. Chlorophyll is almost too well known nowadays, thanks to the advertisements for breath-fresheners, to need explanation: it is the green pigment found in nearly all plants except the fungi.

The absence of chlorophyll from the fungi is important, for it has a profound effect on their mode of life. The most remarkable feature of plants as a group is that they can live without organic food; this is how they differ from animals, which must eat food which was once living if they are to survive. Plants need no external supply of organic matter because they manufacture their own organic food from carbon dioxide in the air; using the energy in sunlight they are able to build up sugar from carbon dioxide and water. Chlorophyll plays an essential part in this chemical miracle, for it traps the energy in light and makes it available for chemical work. Without light and chlorophyll a plant would starve.

Since fungi have no chlorophyll they lack the power to make sugar from air; they therefore have to get their organic food in other ways. In this respect they are like animals. There are two sources of supply open to them. Some fungi make use of dead organic matter of one kind or another, of which there is always a plentiful supply ready to hand in nature: the dead bodies of animals and plants, decaying leaves, humus in the soil, the dung of animals, and the never ending stream of refuse arising out of the activities of man, from plate scrapings to old boots, discarded fabrics, and paper bags. All these things and many others are a ready source of organic food for tens of thousands of different species of fungi, and fungi that feed in this way are called saprobes. The common mushroom, for instance, feeds on organic matter in the soil, and moulds, varying in colour from black or green, through yellow, to almost white, can be seen anywhere where it is damp enough for them to grow.

Other fungi live on organic matter which they obtain from the tissues of a living host: usually a plant, but animals are not exempt from attack. Fungi that do this are called parasites. Some of the parasitic fungi do an immense amount of damage to cultivated plants, and cost the nations of the world hundreds of millions of pounds a year, not only in lost crops but also in fungicide sprays to combat them. Black rust of wheat, Dutch elm disease, brown rot of apples, potato blight, and the mildews that attack a variety of plants from roses to peas, are familiar examples.

Some parasitic fungi are strict in their feeding habits, being able to exist only on their living hosts; such fungi are called obligate parasites. Their number is comparatively few, and is becoming smaller as new methods are found to tempt them to live on culture media containing non-living organic matter. Most parasites possess some capability of living as saprobes if they must, and these are known as facultative saprobes. Various species of *Pythium*, for instance, live as saprobes in the soil for most of the time, and yet can become destructive parasites if presented with a tempting morsel such as a collection of juicy seedlings growing under over-crowded conditions. All grades of parasitism are known among the fungi, from strict parasites down to saprobes which are feebly parasitic on occasion.

Parasitic fungi also vary in the extent to which they are committed to one particular host. At one end of the scale, some parasites will attack only one species of plant, or even one variety within a species, while others can attack a wide variety of different species of host plants. The specialisation of a fungus to one, or a few, species of host is called host specificity, and is seen particularly well in the rust fungi; thus, the black rust of wheat, *Puccinia graminis tritici,* attacks only wheat. On the other hand, most species of *Pythium* are scarcely host-specific at all, being able to ravage almost any seedlings that are small enough for them to attack.

It is not only the parasitic fungi that cause damage, for some of the saprobes are equally pestilential. The wood-rotting fungi are a group causing untold havoc, not only in timber yards, but, much more serious, in buildings. The dry rot fungus, *Merulius lachrymans,* costs us millions a year, and besides the dry rot fungus we have the cellar fungus, the pit prop fungus, and all the other fungi that cause white or brown rots of timber. As for the moulds, nobody could even estimate the cost of the damage they cause, especially in the tropics, where the combination of wet climate and heat provides ideal conditions for their growth.

Let no one think from this that all the fungi are pests, for some rank among the most important of our friends. Apart from the delight of eating mushrooms and other edible toadstools, yeast provides us with bread, wine, and beer, and *Penicillium* has, in recent years, saved many lives and founded the vast antibiotics industry through the discovery of penicillin. Many cheeses, such as Stilton and Gorgonzola, owe their distinctive properties to species of fungi, and fungi are also used in the manufacture of many important substances by fermentation, the most notable being citric acid, which is produced by the fermentation of sugar by the black mould *Aspergillus niger*. Most important of all, the fungi assist the bacteria in the breakdown of organic matter in the soil, a process without which life on earth would be impossible. By and large, the fungi, with all their faults, are well worth while from the anthropocentric point of view.

It is a chracteristic of all the fungi except a few of the most primitive microscopic species that they are composed entirely of extremely fine threads called hyphae. These hyphae may be woven together to form a solid structure, such as the fruit body of a mushroom, and in some instances these may be of considerable size; the largest on record is a specimen of the giant puffball (*Calvatia gigantea*) from America, which is reputed to have been over five feet in diameter. The bracket fungus *Polyporus giganteus* may sometimes measure 60 cm. (2 feet) or more in width. Most fungal structures, however, are smaller than this, and some of the more primitive fungi may consist of single cells that are no more than 10μ across (the Greek letter μ is the microscopist's unit of measurement, and is equal to one thousandth of a millimetre, or one twenty-five-thousandth of an inch). In spite of their small size, these microscopic cells have all the attributes that go to make up a fungus.

In most fungi the hyphae are not woven tightly together except in the formation of special structures such as fruit bodies, but are in the form of loose weft. The hyphae that compose the weft, taken together, are called the mycelium of the fungus.

In most fungi the hyphae are regularly divided into separate cells by cross partitions called septa. The lower fungi, however, are without septa, their hyphae being in the form of uninterrupted tubes throughout their length. An exception to this rule is where reproductive organs are formed; these are usually cut off by septa from the rest of the hypha.

Fungal hyphae contain protoplasm, the essential living substance of which all organisms are composed. The protoplasm contains nuclei, which are minute bodies that exert a controlling influence on the activities of cells; they house the chromosomes, the bodies that carry the hereditary information essential to the proper development of the cell. Most plant cells have only one nucleus per cell, but the cells of many of the fungi are exceptional in containing several nuclei, which appear to share in the work of controlling cell activities. An exception to this is found in some powdery mildews (Erysiphales), which have only one nucleus per cell.

The flowering plants, and their allies the conifers such as the pine, reproduce by means of seeds, but the fungi, in common with other lower plants such as ferns, mosses, and seaweeds, reproduce more simply by means of spores. A seed is really quite a complex structure containing an embryo plant, complete with seed leaves, stem rudiment, and a nascent root. A spore is much simpler, consisting of a minute blob of undifferentiated protoplasm surrounded by a protective wall. A seed must be relatively large to contain a complete embryo plant; even the seeds of orchids, which are the smallest known, are visible to the naked eye. By contrast, most spores are microscopic, some of the smallest measuring less than 2μ in diameter.

Under suitable conditions of temperature, moisture, and aeration a fungus spore will germinate and a new mycelium will grow from it. Perhaps 'germinate' is not the right word to use for a spore, which contains no 'germ', or embryo; but germination is the word universally used by mycologists to describe the initial growth of a spore. When a spore germinates a germ hypha grows out from it; presently the hypha branches, the branching being repeated, and so a new mycelium of the fungus arises. In spite of its minute size, a spore contains in the chromosomes of its nuclei all the necessary information to ensure that the mycelium that arises from it has the characteristic form of the species : you will never find a spore of *Penicillium*, for instance, germinating to form a mycelium of *Agaricus*. This is fortunate for mycologists, who commonly grow fungi in pure culture from generation to generation by the transference of spores from an old culture to a fresh batch of sterile culture medium.

Spores of fungi are of two kinds, sexual and asexual, according to whether their formation has or has not been preceded by a sexual process involving the fusion of two nuclei from the same or from different mycelia. Within these two classes the spores can have various forms according to the way they are produced. Since I shall have repeatedly to refer to these different kinds of spores in describing the fungi, the following account of the more important types may be useful as a guide.

ASEXUAL SPORES

Zoöspores. These are formed only by the lower fungi that grow in aquatic or damp environments. A zoöspore consists of a small cell bearing either one or two fine protoplasmic 'tails', called flagella, by the waving movement of which the zoöspore is able to swim. If the zoöspore has a single flagellum it is nearly always mounted at the back or posterior end of the spore (Fig. 1b), but in one class of fungi, the Hyphochytridiomycetes, the single flagellum is at the front (Fig. 1a), or anterior end. Where the zoöspore has two flagella, they again may be mounted in two ways. Some biflagellate zoöspores are pear shaped, with the two flagella at the pointed (front) end (Fig. 1c), while others are bean shaped, with the flagella inserted at the side in a little groove (Fig. 1d).

The flagella that propel the zoöspores are of two different kinds. One type is branched, giving it a feathery appearance when photographed under the high magnifications produced by the electron microscope. This type is called a tinsel flagellum. The other kind, known as a whiplash flagellum, is unbranched (Fig 1). If only one flagellum is present it is of the tinsel type if inserted at the front end of the spore, whereas if it is at the back end it is always of the whiplash type.

When a zoöspore has two flagella, one points forwards when the zoöspore is swimming while the other points backwards; this applies whether the spore is pear-shaped with apical flagella or bean-shaped with lateral flagella. The forwardly-directed flagellum is always of the tinsel type, while the backwardly-directed one is always a whiplash flagellum.

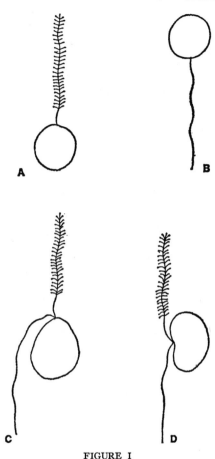

FIGURE I

*Diagram showing the types of flagellation in different kinds
of zoöspores. A, single anterior tinsel flagellum; B, single
posterior whiplash flagellum; C, pear-shaped zoöspore with
two anterior flagella; D, bean-shaped zoöspore with two
lateral flagella. All greatly magnified.*

Zoöspores are formed, usually in considerable numbers, in cells
called zoösporangia. In primitive unicellular fungi the single cell
forming the body of the fungus may also serve as a
zoösporangium, but in most fungi in which zoöspores are produced the zoösporangia are special cells carried on the hyphae
that compose the mycelium of the fungus. The zoösporangia may

be of any shape; commonly they are spherical, ovoid, or cigar-shaped.

Aplanospores. Any spore that is without flagella can be called an aplanospore. There are several different kinds.

Sporangiospores. A sporangiospore is a spore that is formed, like a zoöspore, in a sporangium. It differs from a zoöspore in having no flagella. Sporangiospores are found in many of the moulds (Mucorales). The sporangium may be stalked, in which case the stalk is called a sporangiophore.

Conidia. Conidia are the characteristic asexual spores of the higher fungi, though some of the lower fungi may also form them. A conidium is not at any time enclosed in a sporangium; it is produced on the end of a hypha called a conidiophore, which may or may not have a special cell called a phialide whose function is to produce a conidium or, more usually, several conidia one after another. One conidiophore may bear several, or even many phialides (Fig. 2). A conidium may consist of a single cell or it may have several cells, and its size is variable; the conidia of some fungi are less than 2μ in diameter, while those of others may be more than a tenth of a millimetre long.

Chlamydospores. The spores that I have so far described are normally intended to germinate as soon as they are mature, though some may have considerable powers of remaining viable for a long time if stored in a dry place. A chlamydospore is usually a resting spore, enabling the fungus to survive a period of bad conditions, such as extreme drought, which might kill the vegetative mycelium. A chlamydospore is formed by part of a hypha accumulating a supply of reserve food and surrounding itself with a wall which is usually thick and relatively impenetrable, giving it a measure of protection against unfavourable conditions.

SEXUAL SPORES

Sexual spores are formed as a result of sexual reproduction in the fungi; they are not a part of the sexual process, but their formation follows the sexual act. Like asexual spores, sexual spores are of several different kinds.

The sexual spores of fungi are named according to how the

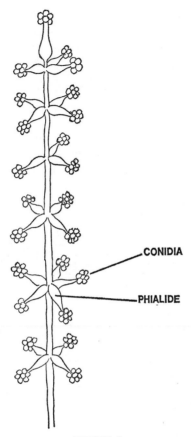

CONIDIA

PHIALIDE

FIGURE 2

Part of a conidiophore of Verticillium *bearing whorls of
phialides with conidia. Greatly magnified.*

sexual reproduction is carried out. Sexual reproduction in any
living organism consists of the fusion of two cells called gametes,
one male and one female. We can distinguish two distinct
phases : plasmogamy, or the fusion of the protoplasm of the two
gametes, and karyogamy, the fusion of their nuclei. It is
karyogamy that is important in the act of fertilisation; usually
the plasmogamy is purely incidental and may be virtually
non-existent.

In many of the lower fungi there is a definite female egg cell which is fertilised by a male gamete; the latter may be a motile flagellated cell or it may be contained in a non-motile structure called the antheridium. When the egg cell has been fertilised it usually becomes a spore called an oöspore. Where, on the other hand, there is no organised egg cell, and the gametes, or the organs containing the gametes, which fuse with one another are very much alike in form, the spore which is formed as a result of the fusion is called a zygospore.

In the higher fungi there are two principal types of sexual spore. In the great group called the Ascomycotina, the sexual process (see Chapter seven) results in the formation of a special kind of sporangium called an ascus, containing spores, usually, though not always, eight in number, called ascospores. In the other major group of higher fungi, the Basidiomycotina, the sexual fusion is followed by the development of a structure called a basidium, on the surface of which basidiospores are formed. There are usually four basidiospores per basidium.

SEX AND THE FUNGI

In most organisms sexual reproduction is clear cut, plasmogamy being followed almost immediately by karyogamy, so that the two phenomena can be clearly seen to be part of the same process. In the higher fungi – and to some extent in the lower fungi too – there is what one can only describe as a casual attitude about sex. This shows itself in a number of ways, one of which is the postponement of karyogamy until a long time after plasmogamy. In the Basidiomycotina, for instance, plasmogamy usually occurs quite early in the life history of the fungus, while karyogamy does not occur until the life cycle is drawing to its close. How this comes about, and the effects it produces, will be described in the next and following chapters.

Some fungi – the Deuteromycotina – do not undergo any apparent sexual process at all, reproducing themselves entirely by means of asexual spores. In many of them, however, this lack of sex is only apparent, for they undergo what is called the parasexual cycle, a complex scheme of plasmogamy and karyogamy which does not involve any special sex organs, or take place at any specified point in the life cycle. Parasexuality in the fungi is of recent discovery, and its complexities are beyond the scope of this book.

HETEROTHALLISM

Heterothallism is of frequent occurrence in fungi of all kinds. It is an extremely complex phenomenon, but stated baldly and, I am afraid, somewhat over-simply, it is the fact that in many fungi sexual reproduction can only occur if two mycelia of opposite mating potentiality are present. Thus, a given species of fungus consists of two strains, often called the plus and minus strains, identical in appearance but differing in their capability of undergoing sexual reproduction each with the other. A plus strain can only mate with a minus strain and not with another plus; similarly a minus strain will mate with a plus strain, but not with a minus. The plus and minus strains are said to be compatible, while two pluses or two minuses are incompatible.

One might think that this was simply a difference of sex, one of a pair of compatible strains being male and the other female, but this is not so. Many fungi have male and female sex organs of recognisably different form borne on the same mycelium. If such a fungus is heterothallic, the male organ is incapable of fertilising the female organ on the same mycelium; for successful fertilisation to take place the male and female organs must come from different, compatible mycelia. Clearly, then, heterothallism differs from sex, though it is also akin to it.

THE NAMES OF FUNGI

A few fungi, such as the common mushroom and the death cap, have 'popular' names, but most of them, since they are not familiar to the man in the street, are known only by their scientific names. They are named under the binomial system invented by the great Swedish biologist Linnaeus, which is universally used for the naming of animals and plants. Each fungus has a name consisting of two Latin words, the first being the name of the genus, followed by the name of the species. The specific name, or epithet, should be descriptive of the fungus, but it seldom is in this imperfect world. Thus, the common mushroom is called *Agaricus campestris*, *Agaricus* being the name of the genus and *campestris* being the specific epithet. *Campestris* means 'appertaining to plains or level country', which I suppose is a fair description of the sort of place where you may expect to find the mushroom growing, though mushrooms are plentiful enough on Cornish hillsides.

The Latin names of fungi are always printed in italics, the generic name beginning with a capital and the specific epithet with a small letter. If written, they should be underlined. It used to be the practice to begin the specific name with a capital letter where it was derived from a proper noun, but this serves no useful purpose except to display the erudition of the writer, and the habit is rapidly dying out, unlamented except by a few die-hards.

The idea of having both a generic and a specific name for every fungus is not merely a convenient means of sorting them out in a card index: its real purpose is to express relationship. Since all living things have arisen by a gradual process of evolution it follows that the degree of relationship between different organisms varies: an organism may be closely related to some of its fellows and less closely related to others. Species of fungi that are closely related to one another are placed in the same genus: thus we have the common or field mushroom (*Agaricus campestris*), the horse mushroom (*A. arvensis*), and the cultivated mushroom (*A. bisporus*), all placed in the genus *Agaricus* because their relationship is believed to be close. This is shown by the fact that they all have purple spores, a ring, and no volva (for an explanation of the term 'ring' and 'volva', see the next chapter). The existence of certain other minor resemblances clinches the relationship.

The idea of relationship is further expressed in the classification of the fungi as we proceed from smaller to larger groups. Related genera are grouped into families, and related families into orders. Thus, the genera *Agaricus* and *Melanophyllum* are placed in the family Agaricaceae (note that the name of a family ends in -aceae and is not printed in italics). Similarly, all the families of fungi which have fruit bodies in the form of a (usually) stalked cap with the spores formed on radiating gills beneath it are placed in the order Agaricales (the name of an order always ends in -ales). Finally, all the orders of fungi in which the sexual spores are basidiospores formed on basidia which are exposed to the air when mature are placed in the class Hymenomycetes, of the sub-division Basidiomycotina.

Within the various major groups, large and small, are sub-groups: we have sub-classes, sub-orders, sub-families, sub-genera, and sub-species or varieties. The classification of the fungi, like that of all plants, is quite a complex affair, and there are specialists called taxonomists whose job it is to look after the classification of organisms, and to try to keep ordinary mortals like you and me on the rails.

There is no need, however, to be depressed at the intricacies of nomenclature and taxonomy. If you can recognise a particular fungus and *know it when you see it again,* you are already on the way to being a mycologist even though you do not know the group to which it belongs, and have no other name for it but 'that peculiar thing which turns red when you break it and is so good to eat'. Moreover, one day, quite soon, you will suddenly find that you can reel off the Latin names of fungi, and the names of the groups to which they belong, without even having to think about it.

2 Mushrooms and Toadstools

Most people are very firmly convinced that there is a fundamental difference between a mushroom and a toadstool; for
one thing, a mushroom is edible while a toadstool is well known
to be deadly poisonous. Nothing could be more wrong, for a
mushroom is only a particular species of toadstool. On the
continent they know better. The French language makes no
distinction between mushrooms and toadstools, the word *champignon* serving for both. In Italian, mushroom soup is *brodo dei
funghi*. In markets all over Europe (except Britain) you will find
many different kinds of edible toadstools offered for sale. Why
the British maintain their insular contempt for toadstools is not
known; possibly it may be an extension of the 'wogs begin
at Calais' habit of thought which was once the main source of
our greatness, but which, in the light of the Common Market,
may now have to be revised.

In actual fact, there are only three species of toadstools that
are *really* deadly. Two of these are rare, but unfortunately
the third is quite common. A few dozen may cause symptoms
ranging from serious illness to slight discomfort if eaten, and
most of the remainder, although completely harmless, vary from
insipid to downright nasty. On the other hand, there are
hundreds of species that are good to eat, and some of them are
as good as, if not superior to, the common mushroom. There is
much to be said for mycophagy as a habit – provided that you
know what species of fungus you are eating with absolute
certainty.

There are many fables concerning the ways in which an
edible toadstool can be distinguished from a poisonous one. All
are untrue, and any would-be mycophagist who places the
slightest reliance on any of them is likely to end up in hospital,
if not in the mortuary. The commonest of these rumours is that
an edible toadstool will 'peel' easily, while a poisonous one will

24

not. This is complete balderdash, and dangerous at that, for the death cap (*Amanita phalloides*), the most deadly of the poisonous toadstools, has a skin that peels off as easily as that of any mushroom. The story that a poisonous toadstool will blacken a sixpence if one is pushed into its flesh is equally fallacious; as far as I am aware, *no* fungus will do this.

Mycophagy is a fascinating and rewarding hobby, for nothing is pleasanter than to go for a walk on a fine October afternoon and to return with a basket of edible toadstools for one's supper as a bonus. To an uninitiate contemplating mycophagy, however, I must give two pieces of advice. One is to eat only those fungi that a properly qualified expert (not merely a 'chap who knows about these things') has told you are harmless. The other is to eat only a little at first, for some unfortunate people are allergic to toadstools that are harmless to others, and it is as well to find out whether you happen to be one of these before gorging yourself or the consequences could be quite serious.

THE COMMON MUSHROOM

The common mushroom (*Agaricus campestris*) is the most familiar of the toadstools: indeed, it is the most familiar of the fungi to most people. It is fairly typical of the kind of structure found in the order Agaricales, though it is a little unconventional in certain matters relating to the behaviour of the nuclei of the cells in its fruit body.

The first thing to realise is that the mushrooms that we see growing are not the whole of the fungus: they are, in fact, only the fruit bodies, sporophores, or basidiocarps – call them which you will – that bear the spores of the fungus. The mycelium of the mushroom is buried in the soil, and consists of a loose network of hyphae just below the surface which produces fruit bodies at intervals. The mycelium nourishes itself on the organic matter or humus contained in the soil, living entirely as a saprobe. There is nothing spectacular about it, so we will concentrate on the fruit body, which offers more possibilities. Incidentally, when you buy mushroom 'spawn' from a horticulturist you are buying compost containing the mycelium of the mushroom, not the spores.

The fruit body of the mushroom consists of two parts. There

is a rounded stalk or stipe, which bears a flat, circular cap or pileus on its end. If you look underneath the cap you will find many vertical plates or gills, arranged like the spokes of a wheel. Some of the gills stretch right from the edge of the cap to the central stalk, some reach only about half way, and some are even shorter, extending for only about a quarter of the way from the edge of the cap. The gills are arranged in such a way that between each pair of long gills there is one of medium length, and each is separated from its neighbour by a short gill. This arrangement avoids overcrowding of the gills near the centre of the cap, an important point when we remember that the spores are carried on the gills and have to fall down between them when they are released.

The mushroom, like all fungi of the order Agaricales and a great many other fungi as well, belongs to the sub-division of the fungi called the Basidiomycotina, in which the characteristic spores are basidiospores borne on structures called basidia. The basidium may take several different forms in the Basidiomycotina, but in the Agaricales, as well as in most other fungi with basidia, it is a club-shaped cell at the end of a hypha. In the Agaricales the hyphae that bear the basidia cover the surface of the gills. The basidia themselves, which are extremely numerous, form the fertile layer or humenium (Plate one).

The young basidium has two nuclei, but before the basidiospores begin to develop these fuse, and the resultant fusion nucleus divides twice, so that the number of nuclei is now four. The fusion of nuclei in the young basidium is the culmination of the sexual act, begun much earlier, as we shall see in a moment.

While the fusion nucleus in the basidium is dividing, four little protrusions appear at the end of the cell. These grow out into four slender horns, called sterigmata (singular, sterigma). The tip of each sterigma then begins to swell, and an oval spore, the basidiospore, is formed. Finally, one of the four nuclei in the basidium passes into each basidiospore.

When fully formed a basidiospore is usually ovoid in shape, and is always attached eccentrically to the tip of the sterigma on the basidium. At one side of the spore, just at the point of attachment to the sterigma, is a minute protrusion called the hilum. When the spore is mature it is shot off the end of the

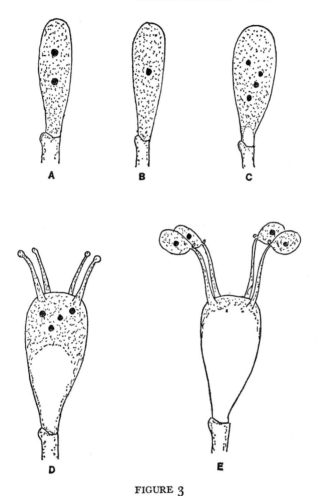

FIGURE 3

Five stages in the development of a basidium and its basidiospores.

sterigma : it does not passively drop off as one might expect. The four basidiospores from each basidium are discharged one after another.

The forcible discharge of the basidiospores is understandable when we consider their method of dispersal. Basically, the basidiospores are dispersed by wind, but before they can become caught up in the air currents they must first fall clear of the protecting gills. These stand vertically, and the force with which the spores are discharged is just about sufficient to carry them away from the surface of the gill and into the space between, so that their fall is unimpeded.

This arrangement will only work if the gills are vertical, so that a spore, falling straight downwards under the influence of gravity, will meet with no obstruction. If the gills are to remain vertical it is essential that the agaric shall grow straight upwards. There are two mechanisms that ensure this. One, which we may call the coarse adjustment, is that the stalk is negatively geotropic: that is, it grows away from the direction of the force of gravity. You can often see this in a toadstool which has started to grow out of the side of a tree stump, for the stalk, which may have started out in a horizontal direction, soon bends upwards.

This bending of the stalk places the cap in a horizontal plane, so that the gills are approximately vertical. Any slight inaccuracy in their positioning is looked after by the second of the two mechanisms, which we may call the fine adjustment. The gills themselves are sensitive to gravity, so that by growing slightly faster on one side than the other they can adjust themselves accurately to the vertical plane.

At the instant before a basidiospore is shot off its sterigma a tiny droplet of fluid appears at the junction between the hilum and the sterigma. This has been generally thought to be a drop of water, but recent research suggests that it is in fact the last droplet of protoplasm from the basidium, the remainder having passed into the spore during its development. When the spore departs it carries the droplet with it.

A word must be said here about the binucleate condition of the young basidium before the nuclear fusion that precedes the formation of the basidiospores. This is not peculiar to the mushroom, but is found in all the Basidiomycotina, and it represents something fundamental in their life histories.

A basidiospore has only one nucleus, a condition that is called monokaryotic. Sometime between the germination of the basidiospore and the formation of the hymenium of the fruit body the cells have to become binucleate (dikaryotic). How this is done has only been investigated in a few of the thousands of different species of toadstools, but the most common method seems to be that found in the ink-caps (*Coprinus*). Here the germination of the basidiospore gives rise to a monokaryotic mycelium, but a pair of monokaryotic mycelia soon join together, so that a dikaryotic cell is produced. From this cell grows a dikaryotic mycelium, from which fruit bodies are formed. Thus, the fruit body is dikaryotic from its inception.

Coprinus is heterothallic. There are two compatible mating strains, plus and minus, and for the establishment of the dikaryon a plus strain must fuse with a minus strain.

Fusion between two monokaryotic mycelia is not necessary in order to establish the dikaryon. In *Coprinus,* and many other Basidiomycotina, the monokaryon may produce tiny spores, called oidia, which resemble conidia in the way they are formed. An oidium may fuse with a cell of a monokaryotic mycelium and, giving up its nucleus, form a dikaryotic cell from which the dikaryotic mycelium arises.

The mushroom is abnormal in that the dikaryon is not established before the formation of the fruit body, for the cells at the base of the fruit body may contain an indefinite number of nuclei, and the dikaryotic condition does not become general until the hymenium is about to be formed. How this happens is still a mystery.

The fusion of the two nuclei in the young basidium is the culmination of sexual reproduction that was begun much earlier. Plasmogamy (fusion of protoplasm) occurred when the dikaryon was established by the fusion of two monokaryotic cells. Instead of plasmogamy being immediately followed by karyogamy (fusion of nuclei) it is postponed until much later; a whole phase in the life history of the fungus, the dikaryophase in which the cells are dikaryotic, is interpolated as it were, between plasmogamy and karyogamy.

Such a postponement of karyogamy until a long time after plasmogamy is rare among living organisms other than fungi, but among the fungi it is common. It reaches its logical conclusion in the Basidiomycotina, where the interpolated dikaryophase occupies most of, or even the whole of, the life history. It is also seen, to a lesser degree, in the Ascomycotina, the other great group of higher fungi, and it is found in some of the lower fungi. Why the fungi should have adopted this curious quirk in their sexual reproduction we do not know, and it is even more curious when we find it foreshadowed in the moulds which, as far as we know, are not all that closely related to the higher fungi. It is hard to see what advantage, if any, the fungi gain by this exceedingly odd behaviour.

OTHER AGARICS

The order Agaricales is a large one, containing some thousands of species all characterised by a hymenium that covers the surface of gills beneath the cap. Many attempts have been made to work out a natural system of classification for them, showing the relationships between the different families, genera, and species. Although each new system is a little nearer the ideal than its predecessor we are still a long way from devising a truly natural system, and the agarics are still mainly classified according to such comparatively artificial characteristics as the colour of their spores, the shapes of their caps, etc.

One might think that to determine the spore colour of an agaric one would only have to glance at the gills, since these are covered with spores, but this is not so, for the gills themselves may be coloured; a species with white spores and brown gills, for instance, might well be mistaken for a brown-spored species. In order to determine accurately the spore colour the spores must be examined by themselves against a white or colourless background. This is usually done by taking a spore print of the fungus. The stalk is cut off close to the cap and the cap is laid, gills downwards, on a sheet of white paper, or better still, glass. After leaving it undisturbed for a few hours we find that the spores, falling from the gills in thousands, have formed a pattern of the gills on the paper. The true colour of the spores can now be seen.

Having determined the colour of the spores the next thing in the identification of a toadstool is the presence or absence of a ring and a volva. To illustrate the meaning of these two terms we will consider the fly agaric, *Amanita muscaria*, an extremely common toadstool that is familiar to everyone, if only from its picture on Christmas cards. It is a fairly large toadstool, white in colour except for the top of its cap, which is a brilliant red, flecked with white scabs.

The fruit body of the fly agaric, like those of most toadstools, begins as a 'button' on the mycelium, the button being ovoid in shape and 2.5 centimetres (1 inch) or so long (Fig. 4a). On cutting open the button you can see the young toadstool, ready to appear, with its stalk and cap tightly folded together. The young fungus is enclosed in a membrane called the universal veil, or velum universale. As the toadstool develops from the button stage the elongation of the stalk ruptures the veil. Part of it is left at the base of the elongating stalk, where it forms a little cup round the base of the stalk. This is called the volva. The rest of the veil is carried upwards and persists as the white scabs on top of the cap.

In addition to the universal veil there is another, the partial veil, or velum partiale, running from about the middle of the stalk to the edge of the cap. As the cap expands this is also ruptured, and its remains are left as a frill round the stalk. This frill is called the ring, or annulus (Fig. 4b).

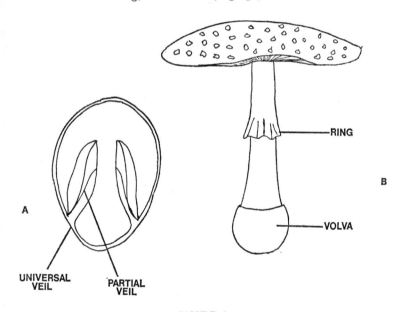

FIGURE 4

Diagram of the formation of the ring and volva in the fly agaric (Amanita muscaria). *A, section of the young fruit body in the "button" stage; B, mature fruit body.*

In the genus *Amanita* both the universal and the partial veils are present and persistent, so that species of *Amanita* have both ring and volva. In some fungi either one or both of the veils may be absent, or so evanescent that they disappear as soon as they are ruptured. Such fungi may have a volva but no ring, a ring but no volva, or they may have neither. The common mushroom, along with all other members of the genus *Agaricus,* has a ring but no volva, though the ring is not always easy to see. The genus *Amanita* is the only *white-spored* genus to have both ring and volva, a point to remember since the death cap belongs to this genus.

It is important to remember that if a toadstool is pulled out of the ground the volva, if present, may be left in the soil; the toadstool then cannot be identified. In gathering agarics to be identified it is necessary to prize the fruit body out of the ground with a knife, so that the volva, if any, will come with it.

The shape of the gills is also important in identifying an agaric. Five conditions are recognised. If the gills are not attached to the stalk they are free (Fig. 5a). If attached to the stalk by part of their width they are adnexed (Fig. 5b), while if they are attached by the whole of their width they are adnate (Fig. 5c). If the bottom edge of the gills is wavy in outline they are sinuate (Fig. 5d), and finally if the inner edges of the gills run down the stalk the gills are decurrent (Fig. 5e).

After the shape of the gills is determined, the shape of the cap is of importance, especially in determining species. It may be flat, convex, concave, pointed, etc. In using this characteristic it should be remembered that the shape of the cap often alters with age. Other points to be considered are the nature of the stalk – whether soft, brittle, or what have you, and, most unreliable character of all in most agarics, the colour. It is amazing how much variability the colour of an agaric can show with age, and the amateur is not helped by books which describe the colour as 'ruffose' or 'livid'!

The identification of agarics is not easy: in fact, of all plants they are the most difficult. Something is now being done with chemical tests, and the use of standard colour charts has been suggested to avoid such colour descriptions as 'ferruginous'. The only course for the beginner is to seek the help of a compe-

Plate 1. Photomicrograph of a section through the gills of a mushroom (*Agaricus campestris*). The edge of each gill is covered with the hymenium of crowded basidia with their spores. Numerous detached basidiospores can be seen between the gills. Greatly magnified

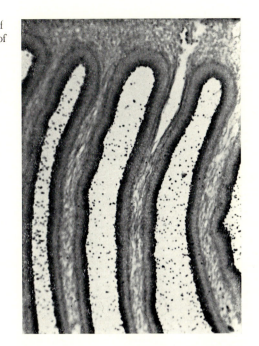

Photomicrograph of a transverse section through the pores of *Polyporus*, showing the hymenium lining the tubes. Greatly magnified Plate 2.

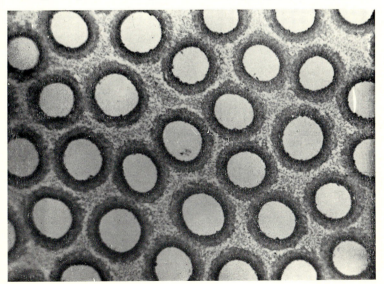

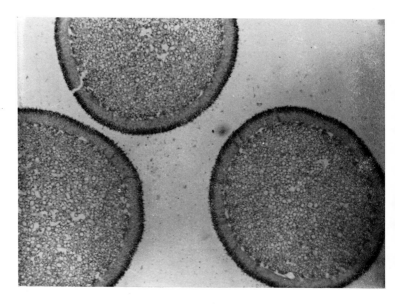

Plate 3. Photomicrograph of a transverse section through three of the spore-bearing spines on the cap of *Hydnum*, showing the hymenium on the outside of the spines. Greatly magnified

Plate 4. Photomicrograph of crushed roots of heather (*Erica* sp.), showing the associated fungus. Somewhat magnified

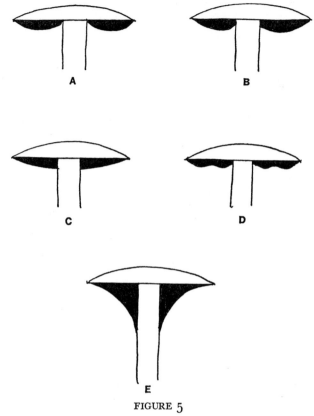

FIGURE 5

*Diagram showing the different forms of gills found in agarics.
A, free; B, adnexed; C, adnate; D, sinuate; E, decurrent.
The gills are shown in black.*

tent mycologist – and even he may be wrong at times. One
important point to remember is that the identification should be
made as soon as possible after the specimen has been gathered,
for once an agaric is removed from the place where it was growing its appearance changes rapidly, and in a few hours it may
even be impossible to say to what species it belongs. If it is a
Coprinus it may even disappear in a flood of inky fluid, for
members of this genus have a habit of rapidly becoming deliquescent when gathered.

SOME COMMON TOADSTOOLS

To attempt to describe even a representative sample of the agarics in a chapter of this length would be impossible, but there are a few common species that demand comment. The fly agaric (*Amanita muscaria*) has already been mentioned, with its bright red cap speckled with the remains of the universal veil. The fly agaric is a poisonous toadstool, and it gets its name from the former use of portions of its cap, chopped up in milk, as a poison bait for flies. The poisonous nature of the fly agaric is due to the presence of an alkaloid. It is said that the poison is not actually deadly to a person in good health, but how good is good? I personally would not risk it. The symptoms include intoxication and delirium, followed by coma, from which the person awakes (if he is lucky) feeling none the worse for his experience. The Norsemen used to chew the fly agaric before battle in order to go berserk; perhaps one day some enterprising Rugby team will issue it to the forwards before a match. The results should be interesting, especially when the stage of coma is reached. It was probably the fly agaric that was in the mind of H. G. Wells when he wrote his delightful short story *The Purple Pileus*.

In the same genus as the fly agaric is the death cap, *Amanita phalloides*. This fungus really is a horror. The flesh contains two poisonous alkaloids, and is fatal even in minute quantity. There is no antidote; to eat it means virtually certain death after a week of indescribable suffering. The death cap is the cause of over ninety per cent of all the deaths from fungus poisoning, usually, one presumes, through people eating it under the impression that it is a mushroom. How anybody outside a mental institution could make such a mistake passes comprehension, for the death cap is really nothing like a mushroom. The colour of the cap is usually a rather bilious yellowish green, though it may on occasion be brownish or white. The stem is usually greenish, with ring and volva conspicuous (Fig. 6a). The gills and spores are white. The death cap is really a woodland fungus, growing in grassy clearings in deciduous woods, but it does on occasion invade grassland bordering on a wood, and I have found it growing among mushrooms. It is usually when it is in the button stage that it is mistaken for a mushroom.

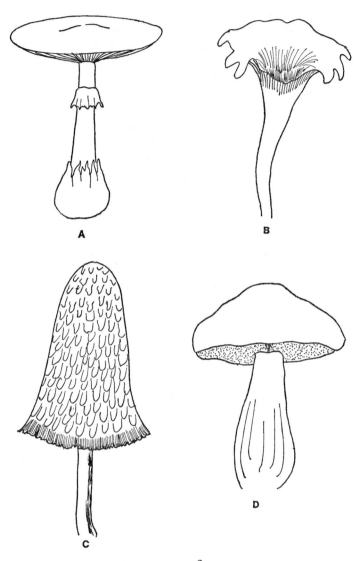

FIGURE 6

Four toadstools of widely differing shape. A, the death cap
(Amanita phalloides); *B, the horn of plenty* (Craterellus
cornucopioides); *C, the shaggy cap* (Coprinus comatus); *D,
the cep* (Boletus edulis).

The destroying angel (*Amanita virosa*) is as deadly as the death cap, which it resembles in a number of ways. It has a somewhat smaller cap than the death cap, and it is entirely white in colour. The presence of a volva, together with the white gills, should clearly distinguish this species from the mushroom.

The blusher (*Amanita rubescens*) is edible. It is a very common fungus which is to be seen during the summer and early autumn in woodlands. It is a large toadstool, with a dull brown or greyish cap up to 15 centimetres (6 inches) in diameter, and covered with greyish, warty blotches; like all species of *Amanita* it has a ring and a volva. Its most notable characteristic, however, and one that distinguishes it immediately from other species of *Amanita,* is the fact that, when either the stalk or the cap is broken, the flesh immediately turns red – hence the name 'blusher'. *A. rubescens* is a very pleasant fungus to eat, with a delicate, slightly astringent flavour when fried. If you wish to eat it, however, make sure that the flesh 'blushes' when broken, for this will distinguish it from the panther (*A. pantherina*), a poisonous species that the blusher somewhat resembles.

The genus *Russula* is an easy one to identify, for all the species have brittle gills that break easily, and the flesh is granular. Unfortunately, the seventy-odd species of *Russula* are extremely difficult to separate, even to the expert. The sickener (*R. emetica*), with its bright red cap, is a common species. The russulas usually have white spores and neither ring nor volva, though in some species the spores are tinged with yellow. Species of *Russula* are usually brightly coloured. None is poisonous when cooked, but many are unpalatable on account of their bitter taste.

Lactarius is another white-spored genus, without ring or volva, which might be mistaken for *Russula* on account of the brittleness of the gills. *Lactarius* can readily be distinguished from *Russula,* however, by the milky fluid that flows from the broken stem; it is on account of this that the various species of *Lactarius* are known as milky-caps. As in *Russula,* no species of *Lactarius* is actually poisonous, but many have an exceedingly unpleasant bitter taste.

The horn of plenty (*Craterellus cornucopioides*) is readily identified by its black, leathery, horn-shaped cap with decurrent gills (Fig. 6b). It is edible, in spite of its appearance, which reminds one of old leather; rather tough, but excellent for flavouring soups. The caps can be dried and kept indefinitely to be used for flavouring when required.

The chanterelle (*Cantharellus cibarius*) is a well known edible fungus that is commonly sold in continental markets. Its cap is at first shaped like a peg-top, but it later becomes concave. The gills are strongly decurrent, and the whole cap is of a livid egg-yellow colour. The fungus smells slightly of apricots when fresh, and this smell becomes very pronounced when the cap is dried, becoming stronger with age.

The honey agaric (*Armillaria mellea*) is a common species which is a destructive parasite on trees and shrubs. The fruit bodies commonly grow crowded together in tufts; this is known as the caespitose habit. The cap is of a dull yellow colour in young specimens, changing to pinkish brown as the fungus ages. *Armillaria mellea* has white spores, and a ring but no volva. When young the fruit bodies are good to eat, but older ones have an unpleasant bitter flavour. *A. mellea* is often found growing on rotting tree stumps in woods.

A remarkable feature of *Armillaria mellea* is that the mycelium is provided with rhizomorphs: long, black strands which look like leather boot laces. These rhizomorphs can grow through the soil for many yards from an infected tree, thus spreading the infection to other trees.

Wood infected with the actively growing mycelium of *Armillaria mellea* is phosphorescent, so that it can be clearly seen in the dark; this was observed long ago by Francis Bacon. The phosphorescence only occurs when the wood is moist.

The mycelium of *Armillaria mellea* forms a mycorrhizal association (see Chapter four) with *Gastrodia elata*, a Japanese leafless orchid, the fungus inhabiting the root of the orchid like a parasite. That the association is not a parasitic one is shown by the fact that the orchid is dependent on the fungus, without which it cannot live.

The sulphur tuft (*Hypholoma fasciculare*) is one of the commonest of fungi and, like *Armillaria mellea,* has a caespitose habit. The fruit bodies are bright yellow, and occur in tufts of up to a dozen on tree stumps. The sulphur tuft has neither ring nor volva, and the spores are purple-black. It is not poisonous, but its bitter taste makes it uneatable.

The inky-caps, genus *Coprinus,* have black spores. When the spores are ripe the cap becomes deliquescent, dissolving away from the edge towards the centre, the fluid produced by the deliquescence of the cap being coloured inky black by the spores; hence the name 'inky-cap'. The genus gets its name *Coprinus* from the fact that many species grow on dung. The caps of species of *Coprinus* have characteristic lines running from the centre to the edge, making them easy to identify. An exception to the coprophilous habit is seen in the shaggy caps (*Coprinus comatus*), a common edible species growing on ground that has fragments of wood buried in it. The shaggy caps is easily recognised by its cap, which is in the form of a narrow cone (Fig. 6c). The fruit body of the shaggy caps may grow to a height of 25 centimetres (10 inches).

The black fluid containing spores that is produced by the deliquescence of the cap has in the past been used as a substitute for Indian ink. It has even been suggested by Boudier that the ink so made should be used for the signing of important documents, on the grounds that the presence of the spores in the ink would, when microscopically examined, immediately distinguish a genuine signature from a forgery.

The true mushrooms (*genus Agaricus*) all have free gills, purple-black spores, and ring but no volva. Three species are commonly eaten in Britain: the common or field mushroom (*Agaricus campestris*), the horse mushroom (*A. arvensis*), and the cultivated mushroom (*A. bisporus*), so called because it only has two spores per basidium. The horse mushroom grows in pastures and is a larger species than the common mushroom, its cap sometimes reaching a diameter of 20 centimetres (8 inches). The horse mushroom smells faintly of aniseed and is extremely good to eat, but care should be taken to distinguish it from the yellow-staining mushroom (*A. xanthoderma*), which can be identified by the bright yellow stain that appears at the base of the stalk when it is cut. The yellow-staining mushroom is harmless to some, but in others it may cause a severe stomach upset. Another yellow-staining mushroom which is best avoided is the wood mushroom (*A. silvicola*), which occurs in woods, especially coniferous woods. The skin of the wood mushroom stains yellow when bruised. On the other hand, the red-staining mushroom, *A. silvaticus,* which also occurs in coniferous woods, is extremely tasty to eat. It is a small mushroom with a reddish brown cap bearing reddish brown scales, and the flesh stains reddish pink if cut when young; this reaction is lost as the cap ages.

FAIRY RINGS

Few people are unfamiliar with the phenomenon known as a fairy ring : a circle of grassland, or on a lawn, where the grass grows taller and a darker green. There are several types of fairy ring, a common one being a double ring of enhanced growth, the two rings being separated by a zone of poor growth, or even bare ground.

In olden times there was no difficulty in 'explaining' the formation of a fairy ring. They were rings made by fairies (or gnomes, or elves, or what have you) dancing in the still hours of the night when all good people were abed. This theory did not lack confirmation, for there was always a bibulous character around who had seen the fairies dancing as he weaved his unsteady way home late at night.

As man became more sophisticated and science began to take the place of fable, other explanations of fairy rings were sought. At first, nobody suggested a biological solution to the problem; this was understandable in an age when science meant physical science. Many theories were propounded, most of them notable for their ingenuity rather than their credibility. One theory that was widely held was that the rings were produced by lightning striking the ground. Finally, about a century and a half ago, the truth was realised : the rings were due to the growth of a fungus.

Many toadstools can form fairy rings, but the most important is the fairy ring champignon, *Marasmius oreades.* This handsome toadstool has a yellowish brown hemispherical cap, becoming flattened with age, but usually retaining a central boss which gave it its old name, Scotch bonnets. The stem is the same colour as the cap, while the gills are yellowish white and free. The fruit bodies are formed round the periphery of the fairy ring, which may be anything from a few metres to fifty metres or more in diameter.

The reason for the formation of a ring is simple, once the cause is known. The fungus starts its life, say from a spore, at a point in a field, and the mycelium grows steadily outwards in all directions, forming a circular patch. With increasing size and age the hyphae in the middle of the circle begin to die off, so that a ring is left, increasing in diameter as the fungus continues to grow. As the new mycelium spreads ever outwards the old mycelium inside the ring dies.

At the edge of the ring the advancing mycelium acts on the organic matter in the soil. Proteins, in particular, are converted into simpler amino-acids, and these are further acted on by soil bacteria, forming salts of ammonia and finally nitrates. The grass benefits from this free dressing of nitrate, and the effect is seen in the increased luxuriance of its growth.

Behind the advancing hyphae the old mycelium of the fungus is dying off, producing the same effect on the grass through the agency of bacteria which break down the proteins in the dead mycelium. This is why two concentric rings are so often found.

We have yet to explain the bare patch between the rings. This is more difficult, and I must confess that botanists are not yet agreed about its cause. The most likely theory at the moment is that in this area the soil is filled with actively growing hyphae of the fungus, which clog the spaces between the soil particles so that both drainage and aeration are impaired. The grass roots are in a state known as physiological drought: though rain may turn the surface of the soil into a quagmire, none can penetrate to the grass roots; as with the Ancient Mariner, it is a case of

> Water, water every where,
> Nor any drop to drink.

With the lack of water the bacteria in the soil are unable to do their work of breaking down organic matter, so that there is a shortage of minerals for the grass roots to absorb. The grass dies from drought and starvation.

If we know the rate of growth of the fungus it is easy to calculate the approximate age of a fairy ring. *Marasmius oreades* grows at a rate varying from 12.5 to 32.5 centimetres in a year; an average figure for a number of years, divided into the radius of the ring, should give the age of the ring. The results of such calculations are a little staggering. Some fairy rings are centuries old; the age of one near Belfort, due to the growth of *Clitocybe geotropa,* is nearly seven hundred years. It must have started growing in the days of Edward I.

Although fairy rings are normally formed only on grassland, if the grasss is ploughed up and the land put down to corn the ring often continues to grow. Its presence in the corn can be detected by the increased height of the corn round the ring, with a poorly grown area just inside it.

THE BOLETACEAE

This is a small but important family of toadstools which differs from other Agaricales in the way in which the hymenium is carried. Whereas in the Agaricales the hymenium covers the surface of gills placed beneath the cap, in the Boletales the lower side of the cap is perforated by a very large number of fine pores, each pore leading to a tube running up into the cap. The hymenium of basidia lines the surfaces of these tubes.

The important genus in the Boletales is *Boletus,* which contains nearly all the species. Many of the species of *Boletus* form mycorrhizal associations with the roots of forest trees (see Chapter four).

The edible boletus or cep (*Boletus edulis*) is, next to the mushroom, the most-eaten fungus in Europe. In France it is known as *cèpe,* and in Germany as *Steinpilz.* The cap is bun-shaped, as are the caps of many species of *Boletus,* with a massive stalk which swells noticeably towards its base (Fig 6d). The flavour of the cep is delicious. Another common edible species is the bay boletus, *B. badius,* which is found in coniferous woods, especially where the soil is sandy. It has lemon yellow tubes which turn blue or bluish green when touched. The flesh immediately turns blue if cut or broken.

No species of *Boletus* is deadly, but the devil's boletus, *B. satanus,* will cause severe vomiting, especially if eaten raw. The cap is thick and almost hemispherical in outline, and can be as much as 20 centimetres (8 inches) across. It is of a whitish grey colour. The pores are yellow at first, but later they become stained with orange and red. As in *B. badius,* the pores and the stalk become blue when touched : the flesh of the cap also turns blue when cut or broken. The devil's boletus has an unpleasant smell, but the taste is mild and pleasant.

Boletus parasiticus is an interesting little toadstool which is parasitic on false truffles, especially *Scleroderma vulgare.* It is a small fungus, from one to two inches across the cap, which is olive-yellow in colour. It is the only parasitic species of Boletus, and is easily recognised by its habitat.

3 Bracket, tooth, club, skin, and jelly fungi

The bracket fungi are so called because their caps form brackets which stand out horizontally, like shelves, from the objects on which the fungus is growing – usually a tree trunk, a dead stump, or some other piece of wood. The true bracket fungi belong to a family called the Polyporaceae (order Aphyllophorales), in which, as in the Boletaceae, gills are replaced by fine vertical pores (Plate 2), but some of the stalkless Agaricales may also be loosely included.

The number of spores produced by one fruit body of a bracket fungus is astronomical, running into thousands of millions. It has been calculated that one fruit body of *Ganoderma applanatum*, a bracket fungus belonging to the Polyporaceae, may shed 30,000,000,000 spores a day, and continue to do so for a whole growing season, the total annual production of spores being in the region of 5,500,000,000,000. Yet of all these billions of spores all but a few fail, for one reason or another, to produce a new mycelium.

THE POLYPORES

The Polyporaceae may be parasites or saprobes. The commonest is *Coriolus versicolor,* a small bracket fungus that grows as a saprobe on stumps and fallen branches of hardwood trees. The bracket is no more than 5 to 7.5 centimeteres (2 to 3 inches) across, and bears on its upper surface a series of concentric zones of different colours, mainly shades of grey and brown. When young it is a beautiful object, but as it gets older the colours fade, the cap becoming a dull brown with concentric zones of paler colour. *C. versicolor* causes a white rot of infected wood.

Another common bracket fungus, this time a parasite, is the birch polypore, *Piptoporus betulinus*. The fruit bodies are confined to birch trees, where they burst through the bark of trunk and branches, often in considerable numbers. The fruit body is shaped rather like a horse's hoof, with incurved margin and white pores; the top of the fruit body is pale brown.

The birch polypore causes a brown rot in the trunk and branches of birch trees, resulting in the death of the tree. It is sometimes called the razor strop fungus from the former use of the fruit bodies for stropping razors. The flesh of the fruit body is extremely absorbent, and there are records of its use as a desk blotter, a fresh surface being formed by slicing off the old one when saturated with ink.

The beef steak fungus, *Fistulina hepatica,* is found on old oak trees, and less frequently on sweet chestnuts. It forms broad, thick brackets which may measure as much as 20 centimetres (8 inches) across. The upper surface of the bracket is soft, and the flesh, though also soft, has a fibrous texture like that of meat. The outside of the bracket varies from dull crimson to the colour of fresh liver, while the flesh is purplish streaked with lines of a paler colour. The pores are cream in young specimens, becoming pink with age and, especially, with handling.

The beef steak fungus attacks oak trees through wounds in the bark, and as it spreads down the trunk it stains the wood a rich brown colour, especially admired by furniture makers. The beef steak fungus is extremely good to eat when young.

The largest British polypore is the giant polypore, *Polyporus giganteus,* the brackets of which may be up to 60 centimetres (2 feet) across. It differs from the polypores so far described in that the bracket has a short stalk attached to one side (the presence of a stalk is characteristic of the genus *Polyporus*). The huge sporophores usually occur in tufts on stumps of beech and other hardwoods. It may sometimes appear on the soil round a decaying beech stump, where it is attached to the dead roots. The top of the sporophore is dark brown with a lighter coloured edge, while the pores are pale cream when young, blackening with age or bruising. The giant polypore is not poisonous, but its toughness makes it unsuitable for eating.

DRY ROT FUNGI

The Meruliaceae are fungi, probably related to the Poly-poraceae, in which the tubes in the fruit body are exceedingly shallow and the hymenium is found on the ridges between the tubes as well as within them. An important member of the Meruliaceae is the dry rot fungus, *Merulius lachrymans.*

A number of different fungi can cause dry rot of timber, and the different species vary in the seriousness of the rots they cause. Of these wood-rotting fungi *Merulius lachrymans* stands out alone as an arch-destroyer of timber, both from the amount of damage it causes and from the diabolical ease with which it spreads from infected to uninfected wood.

Wood attacked by *Merulius lachrymans* cracks both along and across the grain, breaking into cubical chunks (Fig. 7).

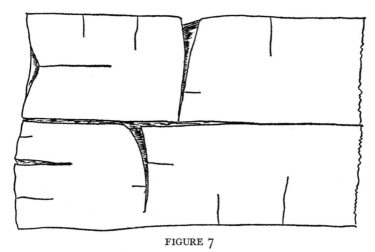

FIGURE 7

Block of wood attacked by Merulius lachrymans, *showing splitting along and across the grain.*

Microscopic examination shows the hyphae of the fungus, well separated from one another, permeating the wood in all directions. On the surface of the wood the mycelium may, under damp conditions, form a fluffy white growth with patches of yellow here and there. On drying, this external mycelium forms a skin which is greyish in colour, with touches of pale purple or lilac.

The external mycelium may give rise to rhizomorphs; these are similar to the rhizomorphs of the honey agaric, but they are stouter, up to one-third of an inch in diameter, and coloured pale grey. The rhizomorphs may extend several metres from the parent mycelium, and they have amazing powers of penetrating through porous solids. They will find their way through mortar and stone walls, and have even been found inside bricks. Wherever they come into contact with damp wood they form a new mycelium, thus carrying the infection far and wide. Brick walls are no barrier to *M. lachrymans;* the rhizomorphs will travel through plaster and brickwork, over concrete and along pipes until they reach wood. It is this that makes *M. lachrymans* so vastly dangerous, for once it has got a hold in a damp corner it will send its rhizomorphs roaming everywhere, infecting sound wood wherever they go.

The name *lachrymans* means 'weeping', and refers to the droplets of water that are often seen on the mycelium, as well as the fruit body, of the dry rot fungus. This is another baleful feature of *Merulius lachrymans*. During its action on wood a great deal of water is formed, and this is carried by the rhizomorphs and enables them to attack wood that is only slightly moist. Most dry rot fungi can only attack wet wood, but *M. lachrymans* has its own private irrigation system and can establish itself in wood that is almost dry.

The rhizomorphs of *Merulius lachrymans* are admirably adapted for their work. They contain three different kinds of hyphae. Some are slender and long, and are similar to those found in infected wood. Others are pointed at the ends and have thick walls; these are clearly for strengthening the rhizomorphs, so that they will not easily be torn. Yet others are wide and thin-walled, and their septa are perforated; placed end to end they form a pipeline to conduct water or dissolved food material along the rhizomorphs, carrying the necessities of life from the parent mycelium to a new one starting up elsewhere. The strengthening and conducting hyphae are not unlike the fibres and vessels that, in flowering plants, serve for strengthening and water conduction, but this resemblance is due to similarity in function, because there is no kinship between a fungus and a flowering plant.

The fruit body of *Merulius lachrymans* is a flattened, fleshy structure, resembling a much-wrinkled pancake; it is cinnamon-brown in colour with white edges. The many wrinkles on its surface run together, forming shallow pores. The hymenium covers the surface of the fruit body, and the number of spores formed is fantastic; it has been said that, if every spore found its target, one fruit body could infect every house in the British Isles. Fortunately the casualties among the spores are high, but plenty survive. Spores of *M. lachrymans* are probably present in every room in every house in Britain, waiting for the fatal damp patch that will enable one or more of them to germinate into disastrous growth.

Dry rot, whether caused by *Merulius lachrymans* or any other fungus, is essentially a result of damp (the words 'dry rot' refer to the effect of the fungus on the wood, and not the condition of the wood when infection takes place). No fungus can attack dry wood, or, for that matter, wood that is submerged in and saturated with water. From time to time some of the wooden piles that supported the old London Bridge eight hundred years ago are fished out of the Thames, unrotted. Wood-rotting fungi, like most other fungi, need air for respiration. Dry rot attacks most readily wood that is damp but not completely saturated, and if ventilation is lacking, so much the better for the fungus, for there is less chance of the wood drying up.

During World War Two many houses were evacuated and left empty. Gutters and down pipes became clogged, windows were broken and let in the rain, and the drop rot fungus throve. The erection of sand-bag blast walls, the covering of ventilators against gas, provided just the conditions that *Merulius lachrymans* and its fellows like best, and they were not slow to take advantage of the situation. Damage by bombing left countless heaps of timber exposed to the rain, and these became foci of infection from which spores in their billions floated into the air, almost as destructive as the enemy bombers. Small wonder that the money spent in combating dry rot in the years immediately following the war was many millions a year.

Wood is the conducting tissue, or xylem, of higher plants. The original cell walls of cellulose are, in wood cells, strengthened by a deposit of hard material called lignin, from which the characteristic properties of wood are derived. Dry rot fungi feed on wood, dissolving the hard material with special catalysts called enzymes that diffuse out of the fungal hyphae and act on the wood; some of these enzymes work by oxidation, using oxygen from the air. As a result of this activity the wood loses its cellular structure and becomes soft and powdery. Dry rot can usually be recognised by pressing the blade of a penknife into the wood: if it enters easily, the presence of dry rot is almost certain, though the cause may not be the dreaded *Merulius lachrymans.*

Several other fungi besides *Merulius lachrymans* can cause dry rot. The cellar fungus, *Coniophora cerebella,* can do much damage in houses if it gets a hold, but it only attacks wood that is really moist. Like *Merulius,* it forms rhizophores that spread the infection from point to point, but these are slender and dark in colour. *Coniophora* does not form as much external mycelium on wood as *Merulius* does, though it may appear in places as a yellowish growth. The fruit bodies of *Coniophora cerebella* are seldom seen in houses, but they may be found on timber stacked in woods; they are brownish, thin, and pimply. Wood rotted by *C. cerebella* tends to crack along the grain rather than across it, a point which will often distinguish it from wood attacked by *Merulius lachrymans.*

Coniophora is a less serious pest than *Merulius,* for several reasons. For one thing, it cannot grow once the wood is dried out. The rhizomorphs are feeble compared with those of *Merulius,* and they lack the built-in water supply system. *Coniophora* is therefore less likely than *Merulius* to spread throughout a building.

A number of other fungi are capable of causing dry rot of timber, among which *Poria vaillantii,* sometimes known as the pit prop fungus, may cause damage to pit props in coal mines. *Poria* is a member of the Polyporaceae.

Some timbers are more resistant to wood-rotting fungi than others, a fact which has been known since ancient times. The timber of the redwood (*Sequoia sempervirens*) is particularly good from this point of view, and is much used for making railway sleepers, fence posts, and other objects that have to remain in intimate contact with the soil. Teak (*Tectonia grandis*) and red cedar (*Thuja plicata*) are also highly resistant to rotting. In many timbers the heartwood is more resistant than the sapwood.

The tooth fungi (family Hydnaceae) are a small but interesting family in which the hymenium is carried on the surfaces of either spines or warts. The spine-bearing genera have fruit bodies resembling the toadstools of the Agaricales, the spines being placed beneath the surface of a stalked cap

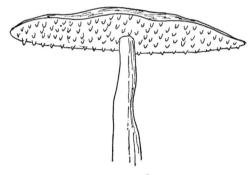

FIGURE 8

A fruit body of Hydnum, *showing the spines beneath the cap.*

(Plate 3 and Fig. 8). The common hydnum (*Hydnum repandum*) occurs in the autumn on the ground among broad-leaved trees. The fruit body varies from pale yellow to buff, with a cap from 5 to 10 centimetres (2 to 4 inches) in diameter with a stalk which is of a paler colour. It is edible, though it should be boiled before frying to remove the slight bitterness that tends to spoil the taste.

Hydnum rufescens is similar to the common hydnum, but the cap is reddish brown. It grows in coniferous woods and, like the common hydnum, is edible. The fir-cone Hydnum (*Hydnum auriscalpium*), a small species with a cap less than 2.5 centimetres (1 inch) in diameter, grows on pine cones; it is readily identified by its habitat.

The members of the Hydnaceae that have warts on which

their hymenia are borne have resupinate (flattened, or skin-like) fruit bodies. Most of them, especially the larger species, are rare, and need microscopical examination in order to identify them.

The club fungi (family Claveriaceae) have club-shaped fruit bodies, the hymenium covering their surfaces. The clubs may be rounded like baseball bats, as in the genus *Clavaria,* or flattened as in *Sparassis.*

Some species of *Clavaria* are extremely common, growing either on rotting tree stumps or on the soil. The fruit body may consist of a single club-shaped structure, or may be branched; the fruit bodies of *Clavaria* often occur in tufts.

The largest British club fungus is *Clavaria pistillaris,* with unbranched clubs from 15 to 30 centimetres (6 to 12 inches) high and 5 centimetres (2 inches) in diameter near the top. The fruit bodies are buff-coloured when young, but become reddish brown with age; they occur on the ground in woods, but are not common. A much more common species is *Clavaria inaequalis,* which has unbranched fruit bodies of a yellowish orange colour, from 5 to 7.5 centimetres (2 to 3 inches) high, produced in groups among grass. *Clavaria cinereus* is another common species, found in woods. The fruit bodies are branched, and of a greyish colour, frequently tinted with purple.

The skin fungi (family Thelephoraceae) usually have a resupinate fruit body, the flattened, skin-like expanse of which is covered by the hymenium. Sometimes, however, the fruit body is stalked, forming a bracket, and resupinate and bracket-like fruit bodies may occur in the same genus, and even in the same species.

Stereum purpureum is the most important species in the Thelephoraceae, because of the damage it causes to fruit trees. The fruit body is of a bright purplish colour; when growing on a horizontal surface it is usually resupinate, but on a vertical surface, such as a tree trunk, it forms small brackets.

Stereum purpureum causes the disease known as silver leaf in rosaceous trees, especially the plum. This disease is characterised by a silvering of the foliage, brought about by the separation of the upper epidermis of the leaf from the palisade layer that lies beneath. The silvering of the leaf, which is due to the reflection of light by air trapped in the space between the two tissues, is not caused by the presence of the fungus in the leaf, but by toxins produced by the fungal growth in the branch, transmitted to the leaf by the water conducting cells.

The commonest member of the Thelephoraceae is *Stereum hirsutum,* which produces a white rot of oak logs. The fruit bodies resemble those of *S. purpureum,* except for the bright yellow colour of the young hymenium, which fades to buff in older specimens. In the bracket form of the fruit body, formed on vertical surfaces only, the top of the bracket is greyish yellow, hairy, and marked with concentric zones of light and dark colour.

THE JELLY FUNGI

The jelly fungi (order Tremellales) have fruit bodies that have the constituency of jelly. They are placed in a sub-class of their own, the Phragmobasidiomycetidae, because they are all distinguished from other Hymenomycetes by having basidia of more than one cell (Fig. 9).

A common species of jelly fungus is the jew's ear fungus (*Auricularia auricula-judoe*). Its brown fruit body is easily recognised by its resemblance in shape and texture to a human ear, and is usually found growing from dead branches of the elder. It gets its somewhat peculiar name from the story of Judas Iscariot, who hanged himself on an elder tree. The fungus was at first called Judas's ear, and the name became degraded, in the course of time, to jew's ear.

The jew's ear fungus is edible, and is particularly good for making soups. It has failed to become popular as an article of food in the western world, probably because of its appearance, but it is much esteemed in China. It sometimes turns up in Chinese restaurants in this country.

The jew's ear fungus belongs to the Auriculariaceae, a family that is characterised by having basidia of four cells, the basidiospores being formed on long sterigmata which carry them up

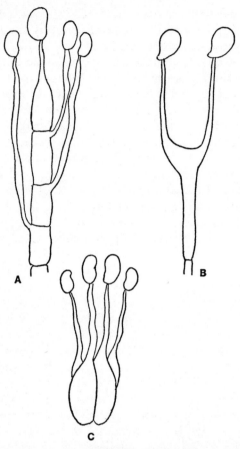

FIGURE 9

Types of basidia found in the jelly fungi. A, Auricularia; *B,*
Dacryomyces; *C,* Tremella. *Greatly magnified.*

beyond the layer of jelly in which the hyphae of the fruit body
are embedded (Fig. 9a).

The fruit bodies of *Dacryomyces deliquescens* are often seen
in wet weather on old fences; when the weather is dry they
shrivel up and almost disappear. They are in the form of small,
hemispherical cushions, about 3 millimetres (1/8 inch) in
diameter, and are of a bright orange or yellow colour. *D. deliquescens* belongs to the family Dacryomycetaceae, the members

of which family all have bisidia with two arms reaching upwards and carrying the basidiospores, of which there are only two, one for each arm. The basidia of the Dacryomycetaceae are often known as 'tuning fork' basidia from their resemblance to tuning forks (Fig. 9b).

The yellow brain fungus (*Tremella mesenterica*) is common on logs and tree stumps at any time of the year. The fruit bodies are bright golden and soft to the touch, with their surfaces thrown into numerous folds, like the convolutions of the human brain. They are from 5 to 7.5 centimetres (2 to 3 inches) across. In *Tremella mesenterica* and other Tremellaceae the basidia are divided into four cells by vertical partitions (Fig. 9c).

Septobasidium (family Septobasidaceae) has a remarkable parasitic relationship with scale insects. The basidiospores of the fungus germinate on the horny covering of the insect and the hyphae enter the insect's body. The insect attaches itself to a host plant, and the fungus forms a mat of hyphae covering the insect. The hyphal mat continues to grow, forming a house in which a whole colony of scale insects lives; some of the insects are parasitised by the fungus while some are not. The insects that are parasitised remain alive in spite of the presence of the fungus, though one of the effects of the fungal parasitism is to make them incapable of breeding.

4 The toadstool and the tree

Toadstools are particularly common in woodlands. This is partly because in woodlands they get the damp, shady environment that most of them prefer, but there is also another reason. Some toadstools grow in partnership with forest trees.

If you dig up some of the shallow feeding roots of the beech you will find, near their tips, a number of short rather thick branches. If a section (a thin slice) of one of these rootlets is examined under a microscope it will be seen that it is covered with a fine weft of fungal hyphae, forming a mantle from 20μ to 30μ thick. The hyphae do not penetrate the root to any great extent, though they can be seen running between the cells of the cortex, or outer layer of the root, where they form what is known as the Hartig net. Here and there hyphae from the mantle round the root run off and connect with a fungal mycelium in the soil.

In spite of appearances to the contrary, the fungus is not a parasite attacking the tree; this is shown by the fact that the beech grows better with the fungus than without it. The fungus and the tree live in partnership, each deriving benefit from its association with the other. This kind of relationship between two organisms, not by any means uncommon, is called symbiosis, a word that means, literally translated, 'living together'. The particular association that exists between certain fungi and the roots of higher plants is known as mycorrhiza ('fungus root').

Since the fungal hyphae are almost entirely outside the root this form of mycorrhizal association is known as ectotrophic mycorrhiza, distinguishing it from other forms of mycorrhiza where the root tissues are more deeply invaded. Ectotrophic mycorrhiza is of common occurrence in forest trees, particularly in the conifers, the birch family (Betulaceae), and the beech family (Fagaceae). In the pine the mycorrhizal rootlets are, like those of the beech, short and stumpy, but they differ from those of the beech in that they are often branched. The mycorrhiza of the pine is often called coralloid mycorrhiza, from the fancied resemblance of the infected roots to coral.

Many different fungi, all of them Basidiomycotina, are able to set up mycorrhizal associations with trees, and many of the toadstools that are commonly found in woodlands are the fruit bodies of these mycorrhizal fungi. Sometimes a particular species of fungus is always associated with a particular species of tree; thus, *Boletus elegans* produces mycorrhiza in the larch (*Larix decidua*) and not, as far as is known, in any other tree. Some fungi, however, are less specific and are able to set up mycorrhizae with a variety of different trees. The most catholic of all the mycorrhizal fungi appears to be *Cenococcum graniforme*, which can form mycorrhizae with fir, spruce, pine, Douglas fir, hemlock spruce, larch, juniper, oak, beech, birch, hazel, hickory, alder, lime, poplar, and willow. Most of the mycorrhizal fungi fall somewhere between these two extremes.

The extent to which the fungus is able to get on without its partner is a matter of dispute, and different fungi appear to vary in this respect. *Boletus bovinus*, for instance, seems to be quite capable of an independent existence in nature; but for most species of mycorrhizal fungi it is at least dubious whether the fungus can live without the tree.

Trees, too, differ in the extent of their dependence on mycorrhizae. Some appear to be able to live quite happily without their fungi, especially if the soil is rich; others seem to depend more on fungal infection in order to live a healthy life. If the soil is poor the need for a fungal partner is greater than if the soil contains plenty of available nutrients.

Although the existence of a state of partnership between tree and fungus has been demonstrated, we still do not know what each partner gets out of the association. It was formerly thought that the fungus, being a saprobe, was able to break down complex organic compounds in the soil, and that some of the proceeds of this were passed on to the tree, which, like most green plants, could not absorb organic food materials by itself. This, however, does not seem to be the case. Most of the mycorrhizal fungi can be grown by themselves in the laboratory on suitable culture media and, when this is done, it is necessary to supply them with sugar as a source of carbon; they cannot break down the more complex carbon compounds that are present. It seems unlikely, therefore, that the tree benefits from fungal activity in breaking down organic matter in the soil.

Much of the water and nutrient minerals absorbed from the soil by the roots of flowering plants enters by the root hairs : minute hairs that cover the surface of the roots near their tips, greatly increasing their absorbing surface. When an ectotrophic mycorrhiza is formed the development of root hairs is largely suppressed, the hyphae of the fungus taking their place. It could be that the hyphae of mycorrhizal fungi are more efficient than root hairs in carrying out this function, so that the absorbing power of the tree roots is increased by mycorrhiza. Phosphates, in particular, appear to be absorbed readily by the hyphae both from the soil and the decaying litter on the floor of the wood.

We are equally ignorant of what the fungus gains from the tree, though it is thought that a supply of sugar is probably the answer. In this connection it is worth noting that young tree seedlings are not attacked by mycorrhizal fungi; it is not until the first leaves have opened and the trees are beginning to manufacture sugar by photosynthesis that the fungus begins its invasion of the roots.

MYCORRHIZA IN THE ERICACEAE

Plants of the heath family (Ericaceae) also show the phenomenon of mycorrhiza, but in a different form from that of forest trees. The Ericaceae includes many common plants of heaths and moorlands, such as the heathers (*Erica* spp.), the ling (*Calluna vulgaris*), and the wortleberries (*Vaccinium* spp.). The roots of these plants, which are extremely fine, are enclosed in a loose mantle of fungal hyphae, somewhat resembling the mantle that surrounds the mycorrhizal roots of trees (Plate 4). In the Ericaceae, however, the hyphae not only enter the cortex of the root freely, but actually penetrate the cortical cells. Because of this internal penetration the kind of association found in the Ericaceae is called endotrophic mycorrhiza.

The hyphae that enter the cells of the cortex remain healthy for a time, but are later digested by the cells. The speed with which this digestion takes place varies according to the season.

In the spring, when the roots are growing vigorously, infection is light, and the hyphae entering the cortical cells are quickly digested. As the year wears on infection becomes heavier, becoming greatest in the autumn.

Some of the early workers on mycorrhiza in the Ericaceae reported finding hyphae of the mycorrhizal fungus in the stem, leaves, and even the seeds of the plant as well as in the roots, and this gave rise to the fable that mycorrhiza in the Ericaceae was a systemic infection : that is, the whole plant was infected by the fungus. We now know that this is not true, at least under normal conditions, though it may possibly occur in abnormal circumstances, as when the resistance of the host to the fungus is lowered by ill health. It seems more likely, however, that the fungus isolated by the early workers from parts of the host other than the root was not the fungus causing mycorrhiza.

There has been much argument over the identity of the fungus that produces mycorrhizae in members of the Ericaceae. It was formerly thought to be *Phoma radicis,* one of the Deuteromycotina, a sub-division of the fungi in which are placed those fungi which, because their sexual reproduction is unknown, cannot be referred to their true taxonomic positions. The results of more recent work, however, makes it unlikely that *P. radicis* is implicated. A fungus has now been isolated from the roots of species of *Vaccinium* that will produce mycorrhizae when inoculated into sterile seedlings. Unfortunately, this fungus does not appear to form spores, at least under the conditions of culture that have been tried, so it cannot be identified. We shall have to await the results of further research before knowing what it is, though we do know that it is not *P. radicis.*

How each partner gains from the mycorrhizal association in the Ericaceae is not known. It was at one time thought that the fungus could 'fix' atmospheric nitrogen : cause nitrogen gas from the atmosphere to combine with other substances and so become available for the nourishment of the host plant. This hypothesis is now known to be false. As in the ectotrophic mycorrhizae of trees, it may be that the fungal hyphae are better at absorbing water or minerals from the soil than the normal root hairs, but we have no proof of this. There is also the possibility that the fungus breaks down organic matter in the soil and thus makes it available to the host plant as food, but again proof is lacking. We are also ignorant of the advantage gained by the fungus; a supply of sugars is a facile suggestion, but one that could well be wrong.

ENDOTROPHIC MYCORRHIZA IN THE ORCHIDS

The orchid family shows a highly specialised form of mycorrhiza which is truly endotrophic in that the mycelium of the fungus is almost entirely within the root. The orchids themselves are among the most specialised of the flowering plants, and differ from the bulk of them in a number of ways. Some orchids are saprobes, having little or no chlorophyll in their leaves, which are reduced to small scales, and obtaining their organic food from outside instead of manufacturing it for themselves by photosynthesis. An important difference from other flowering plants lies in the structure of their seeds.

The seeds of orchids are extremely small, and are produced in prodigious numbers. The seed is drastically simplified. In most flowering plants the seed contains an embryo plant, complete with rudimentary stem and root and either one or two seed leaves or cotyledons. In the orchids the embryo is simply a spindle-shaped mass of cells, without appendages; in exceptional cases the number of cells in the embryo may be as small as eight, and it is seldom more than a hundred.

Germination in an orchid seed begins with the rupturing of the seed coat by the swelling of the seed contents as it absorbs water. Using the small amount of food material stored in the seed the embryo develops into a protocorm: a small swollen structure somewhat resembling a crocus corm in miniature. There growth ceases, and is not resumed until the young protocorm is infected with the appropriate mycorrhizal fungus, a process that may take two years in some instances. Growth and development are then resumed, the tiny plant being fed on organic matter absorbed from the soil by the fungus. Green leaves are formed, and the young plant begins to feed itself by photosynthesis.

It is important to remember that *all* orchids are saprobes during the early part of their growth. Most orchids ultimately develop chlorophyll in their leaves and turn from a saprobic life to photosynthesis, but some species remain saprobes for the whole of their lives. A fairly common example of a saprobic orchid is the bird's-nest orchid (*Neottia nidus-avis*), an orchid with a tangle of roots resembling a bird's nest that is found in beechwoods.

The roots of all orchids are infected by a mycorrhizal fungus. One might reasonably think that infection of the roots takes place from the infected protocorm, but it does not; the roots have to be infected from the soil. The root tubers that are formed in most orchids are also not infected, and the young roots that grow out of them each spring must again get their mycorrhizal fungus from the soil.

The mycorrhizal fungus invades the cortical cells of the orchid root freely, forming coils of hyphae inside them. Later the hyphal coils are digested by the cells that contain them, leaving a structureless mass of indigestible material in the cells. The digestion may take place in any cells of the cortex, but often there is a definite digestive layer in which the fungus is destroyed (Plate 5).

Several different fungi have been identified as causing mycorrhizae in orchids. Two groups are recognisable. On green orchids the fungus is usually a species of *Rhizoctonia,* a fungus which, like *Phoma,* is placed in the Deuteromycotina because of its failure to undergo sexual reproduction. In saprobic orchids such as *Neottia nidus-avis,* on the other hand, the fungus is nearly always a member of the toadstool-forming Basidiomycotina, such as *Marasmius coniatus* and *Xerotus javanicus.* Mention has already been made of the honey agaric, *Armillaria mellea,* producing mycorrhizae in orchids of the genus *Gastrodia.*

It has been shown that the fungi which are responsible for mycorrhiza formation in the orchids are capable of using complex organic sources of carbon; this is only to be expected when one considers the part they play in the life of the young orchid seedling, and the adult saprobic orchids. There seems to be little reason to doubt that the part they play in the mycorrhizal partnership is the breaking down of complex organic matter in the soil, making it available for the orchid to use as food.

We do not yet know what the fungi get out of their relationship with orchids, but the most probable answer is food of some kind or another. The relationship between an orchid plant and its mycorrhizal fungus is a delicate one, and could well be described as a sort of balanced parasitism. The orchid allows the fungus to go just so far, and when the fungus oversteps the mark the orchid defends itself by digesting the fungal hyphae. As long as the parasitism is kept within reasonable bounds both plants benefit from the association, but if either plant gains control of the other they must both die.

VESICULAR-ARBUSCULAR MYCORRHIZA

In recent years attention has been drawn to another form of mycorrhiza that is extremely common among the flowering plants. This is called vesicular-arbuscular mycorrhiza because the hyphae of the fungus frequently bear globular swellings (vesicles), and because minute processes (arbuscles) which are finely branched in the manner of a tree or bush are introduced into the host cells. Although vesicular-arbuscular mycorrhiza has only been discovered fairly recently it is surprisingly common, nearly every family of flowering plants that has so far been examined for this type of mycorrhiza having one or more genera that show the phenomenon. It is known to occur in a variety of important crop plants, including wheat, barley, oats, rye, maize, and clover.

Although it has not been known for long, vesicular-arbuscular mycorrhiza has been in existence for hundreds of millions of years. Fossils of *Asteroxylon mackei*, a plant from the Devonian period that lived more than 350 million years ago, show clearly the presence of vesicular-arbuscular mycorrhiza. The ancient fungus that gave rise to the condition has been called *Palaeomyces asteroxyli*.

As yet we know nothing about the host-fungus relations in this type of mycorrhiza. The fungi concerned have hyphae with no septa, or cross partitions, dividing their hyphae into cells, so thay may be presumed to belong to the lower fungi, but only in relatively few instances have they been positively identified. So far, the three genera of fungi implicated with certainty in vesicular-arbuscular mycorrhiza are *Endogone, Pythium,* and *Rhizophagus. Endogone* and *Pythium* are both well known genera of soil fungi, but *Rhizophagus* is rather a "genus of convenience" designed to cover a number of fungi causing vesicular-arbuscular mycorrhizae which, when they are better known, will probably be resolved into several different genera.

5 Puff-balls, stinkhorns, and bird's nest fungi

The puff-balls belong to the class of fungi called the Gastero-mycetes, which have a type of fruit body which is fundamentally different from anything that I have so far described. In the Hymenomycetes the basidia are fully exposed even when quite young; they may be tucked away on gills under the cap as in the toadstools, or hidden in pores on the lower side of a bracket as in the Polyperaceae, but they are never enclosed in a chamber without an opening. In the Gasteromycetes the hymenium of basidia lines the inside of a closed fruit body, and is not exposed until the basidiospores are ripe, by which time the actual basidia have usually disintegrated. Some of the Gasteromycetes go further than this, for their spores are kept permanently inside the fruit body until it either decays or is eaten by an animal. The name Gasteromycetes means 'stomach fungi', and refers to this enclosure of the hymenium within a cavity.

Another distinction between the Gasteromycetes and the Hymenomycetes is that in the Gasteromycetes the basidiospores are not violently discharged from the basidia. This is reasonable, for there would be no point in shooting off the spores into an enclosed cavity.

THE PUFF-BALLS

The puff-balls (Order Lycoperdales) can easily be recognised by their globular or egg-shaped fruit bodies, which are white or brownish in colour and may be anything from the size of a golf ball to (exceptionally) the size of a football. They commonly grow in grassland and sometimes on lawns or in woods; some species are found growing on tree stumps.

The common puff-ball (*Lycoperdon perlatum*) is widespread throughout Britain, the fruit bodies occurring in grassy places during summer and autumn. The fruit bodies are from 2.5 to 5 centimetres (1 to 2 inches) in diameter and from 7.5 to 10 centimetres (3 to 4 inches) high (Plate 6). When young they are pure white, but as they grow older a brownish colour appears. They are top-shaped, the broad upper portion narrowing towards the base.

The fruit body of the puff-ball is covered with an outer skin, or peridium, of two layers. Inside this is a soft mass of hyphae forming the gleba, in which are situated numerous cavities, the glebal cavities, where the basidia occur; each glebal cavity is lined with a hymenium of basidia (Plate 7).

When the basidiospores are ripe they fall off the basidia into the glebal cavity, which finally becomes full of spores. The hyphae composing the gleba disintegrate, so that finally we are left with the hollow peridium containing a mass of spore powder. Meanwhile, the outer layer of the peridium gradually disintegrates under the action of the weather, leaving the inner layer intact. The inner layer of the peridium has a small pore at its tip through which the spores are puffed out (hence the name puff-ball) at the slightest pressure, such as the impact of a drop of rain. In this way the spores are dispersed. If a puff-ball is gently squeezed the puff of spores can be clearly seen.

The fruit bodies of the puff-balls are good to eat when young.

In the genus *Calvatia* there is no pore for the emission of spores. Both layers of the peridium are fragile and the spores are set free when they fragment. The giant puff-ball (*Calvatia gigantea*), an American species, is said to have reached a diameter of 1.7 metres (5 feet).

The earth stars (family Geastraceae) are close relatives of the puff-balls, and also belong to the Lycoperdales. The best known genus is *Geastrum*, in which the outer peridium splits into several lobes which, when wet, open out on the form of a star, leaving the intact inner layer of the peridium in the centre, like a small puff-ball. The inner peridium opens at the apex by a pore as in *Lycoperdon*. Species of *Geastrum* are usually found on sandy soils, but they are not very common. *G. triplex* occurs in beechwoods.

THE EARTH BALLS

The earth balls (order Sclerodermatales) are not unlike small puff-balls, but they differ from the Lycoperdales in a number of ways. The peridium is thick and hard, and the gleba inside it is usually dark coloured.

The common earth ball (*Scleroderma aurantium*) grows on the ground in woods and on heaths. The fruit body is bun-shaped, with a diameter of from 5 to 7.5 centimetres (2 to 3 inches). The skin is yellowish brown and covered with warts. There is no ostiole for the release of the spores.

Species of *Scleroderma* are sometimes known as false truffles on account of their superficial resemblance to the true truffles. They are sometimes sold to the unsuspecting as truffles; this practice is much to be deplored, for not only are they relatively insipid in flavour, but if eaten in any quantity they are poisonous. Microscopical examination of the spores will distinguish false from true truffles instantly. The spores of *Scleroderma* are minute, no more than 10μ in diameter, with net-like markings on their walls, while those of the true truffles are many times larger and have smooth walls. The true truffles belong to the Ascomycotina, an entirely different sub-division of the fungi from the Basidiomycotina.

THE STINKHORNS

The stinkhorns (Order Phallales) get their popular name from the foetid odour of the gleba, combined with the horn-like appearance of the mature fruit body. The origin of the scientific name will be obvious to anyone who has seen a stinkhorn.

The common stinkhorn (*Phallus impudicus*) is widespread in Britain, occurring in the litter on the floor of woodlands, and in waste places. The mature fruit body consists of a white stalk, from 10 to 12.5 centimetres (4 to 5 inches) high, with a conical top covered with an olive green slimy substance with a smell that is loathsome beyond belief. The stalk springs from a whitish 'cup' resembling the volva in the Agaricales (Fig. 10).

To understand how the fruit body of the stinkhorn fits into the Gasteromycetes one must look at its development. It begins as a small egg from 2.5 to 5 centimetres (1 to 2 inches) long, contain-

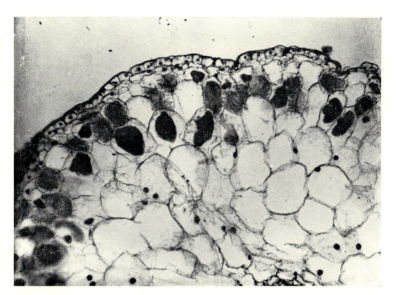

Plate 5. Photomicrograph of part of the cortex of an orchid root, showing the mycorrhizal fungus in the cortical cells. In the outer part of the cortex the coiled hyphae of the fungus can be clearly seen, while just inside the cells contain dark masses of partly digested hyphae. Magnified

Plate 6. The common puff-ball (*Lycoperdon perlatum*)

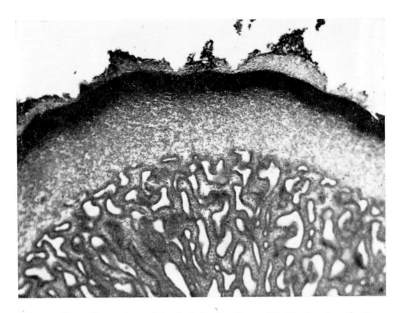

Plate 7. Part of a section of the fruit body of a puff-ball, showing the two-layered peridium and the gleba with its glebal cavities lined with basidia. Magnified

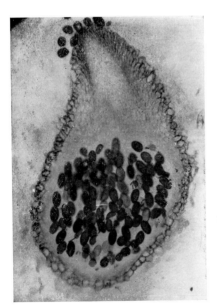

Plate 8. A perithecium of *Sordaria fimicola*, showing the ascospores. The asci, which are thin and transparent, do not show in the photograph. Magnified

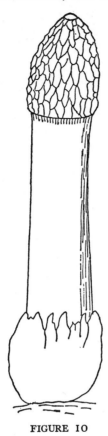

FIGURE 10

Fruit body of the common stinkhorn (Phallus impudicus).

ing a white core which is the future stalk, with the olive green gleba at one end. When the spores ripen the stalk lengthens extremely rapidly, tearing its way through the peridium, which forms the 'shell' of the egg. The remains of the peridium are left behind, enclosing the base of the stalk and forming the 'volva'. The upper end of the stalk, known as the receptacle, bears the basidiospores entangled with the brownish green, evil-smelling, semi-fluid mess that is the remains of the gleba. The whole process of expansion of the fruit body is extremely rapid and may occupy no more than an hour and a half from the first rupturing of the peridium surrounding the egg.

C

The spores of the stinkhorn are disseminated by flies, whose taste in odours differs markedly from ours. Attracted by the foetid smell they crowd upon the newly erected fruit bodies and carry away the spores on their legs and mouth parts, as well as in their guts, from which the spores are voided without harm.

The eggs of the stinkhorn are edible, before the revolting smell develops. The sticky material on the receptacle is said to taste sweet; one salutes the unknown hero who first made this observation.

The dog's stinkhorn (*Mutinus caninus*) resembles the common stinkhorn in all essential particulars, but it is smaller and its smell is much less feculent. It is fairly common among leaves on the ground in woodlands. Its stalk is white or pinkish, with an orange receptacle.

Clathrus cancellatus, sometimes called the lattice fungus, is rare, but it is occasionally found in gardens. It is as beautiful as it is evil-smelling. It develops from an egg in the same way as *Phallus*, but the receptacle hangs from the top of the stalk in the form of a lattice work, the gleba covering the inner surface. The lattice is pinkish red, with a slimy, olive-brown gleba inside it.

THE BIRD'S NEST FUNGI

The bird's nest fungi (order Nidulariales) are a small but remarkable group of fungi whose fruit bodies somewhat resemble minute bird's nests full of eggs. The 'nests' are no more than 6 to 12 millimetres ($\frac{1}{4}$ to $\frac{1}{2}$ inch) in diameter, and are made up of the peridium of the fruit body, split open and exposing the gleba. The glebal cavities, instead of being enclosed in a single mass of gleba, are separate, each with its portion of gleba surrounding it; these form the 'eggs' sitting in the 'nest'. In some genera the 'eggs' are attached to the inside of the 'nest' by a slender filament, the funicle, which plays a part in the dispersal of the spores. Each 'egg' is called a peridiole.

Until fairly recently it was not known how the spores of the Nidulariales were dispersed. It was clear that the peridioles, full of spores, were shot out of the cups, for they had been found clinging to vegetation as much as a yard away from the nearest fruit body. How they got there was a mystery until, about twenty years ago, Buller and Brody showed that the cups of the Nidulariales were in fact 'splash cups'. In rainy weather the cups soon filled with water, and a direct hit by a raindrop produced sufficient turbulence in the water already in the cup to shoot out a peridiole to a distance of a yard or more.

In those genera of the Nidulariales where the peridioles were attached to the cup by funicles the force with which they are ejected is sufficient to break the connexion. The funicles, when wet, expand greatly, trailing behind the flying peridioles like the tail of a comet. At the extreme tip of the funicle a sticky portion, called the hapteron, sticks to any stem or leaf with which it comes into contact. This sharply arrests the peridiole, which is either left hanging by its funicle or, more probably, the funicle winds itself round the object to which the hapteron is attached.

Those genera in which there is no funicle have their peridioles coated with a sticky substance.

FUNGUS ARTILLERY

Included in the Nidulariales is a curious fungus called *Sphaerobolus*, also often known as the fungus artillery on account of its somewhat dramatic ballistic spore dispersal mechanism. *Sphaerobolus* forms a small fruit body less than 3 millimetres (⅛ inch) in diameter. The peridium of the fruit body splits open at the top, exposing the rounded upper surface of the gleba. When the split has occurred the tiny glebal ball is violently shot into the air, often to a height of a yard or more.

The spore howitzer of Sphaerobolus is quite simple, once you know the trick. The peridium is constructed of no less than six layers. When the top of the peridium splits open, revealing the gleba, the lower part of the peridium, below the gleba, separates into two cups, sitting one inside the other. The outer cup consists of three of the six layers of the peridium, while the inner cup is made up of two layers. The sixth layer – the innermost one, in actual contact with the gleba – has broken down, forming a kind of lubricating fluid submerging the gleba, which is now quite free of attachments and sits loosely in the inner cup.

The two layers of the inner cup are made up of quite different kinds of cells. The inner layer (next to the gleba) consists of large, thin-walled cells, forming what is sometimes called the palisade layer, while the outer layer consists of fine hyphae which are firmly interwoven with one another. The cells of the palisade absorb water from the fluid in the cup, so that they tend to increase in size, a tendency which is resisted by the outer layer of fine hyphae. Considerable strain is set up; this is suddenly released in the only way it can be released – by the inner cup turning inside out, giving the palisade cells room in which to swell. In the sudden eversion of the cup the gleba is shot violently upwards (Fig. 11), and that is that.

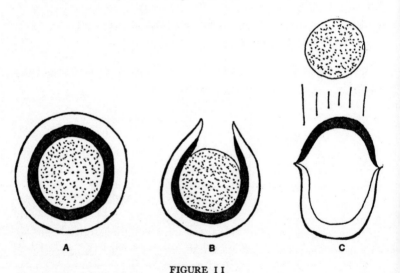

A   B   C

FIGURE 11

Diagram showing three stages in the discharge of the 'spore howitzer' of Sphaerobolus. *Magnified.*

6 The rusts and smuts

So far, with the exception of the dry rot fungi and a few parasitic polypores, we have dealt with fungi in their more friendly mood; even the stinkhorn, vile though it may smell, harms nobody and is even good to eat when young. One cannot, unfortunately, say the same for all fungi. The damage done to crops, to say nothing of garden plants, by parasitic fungi is almost incalculable, and the financial burden we bear on account of their baleful activities runs into thousands of millions of pounds a year. Among these fungal enemies the rust and smut fungi take a high place.

BLACK RUST OF WHEAT

Black stem rust of wheat occurs all over the world and does immense damage to wheat crops every year. The first sign that a wheat crop is infected is the appearance on the stems and leaves of small streaks of a rusty red colour. The individual streaks never reach any great size, but they are formed in such large numbers that the whole crop may appear to be rusting away. The rusty patches are dry and powdery, and after walking through an infected field one's clothes may be streaked with the red powder that is easily shaken off the rusty plants. Later in the season the colour of the rusty patches turns from red to black; hence the name black rust.

The stem rust of wheat is due to a fungus called *Puccinia graminis*. The red patches on the wheat plants mark the places where the red spores, called uredospores, of the fungus are bursting through the epidermis of the host, ready to be carried by the wind to infect another plant. The uredospores, formed in countless millions, rapidly spread the disease throughout a field of wheat. What is more important, they may be carried for miles by air currents and thus spread the infection to another crop.

As autumn approaches the rusty red colour of the spore-bearing patches, or sori as they are called, changes to dark brown. This is because the uredospores are giving place to another kind of spore, called a teleutospore, which is two-celled and coloured dark brown instead of red. This marks the end of the season, both for the wheat and for the fungus. The teleutospores are thick-walled spores, designed to withstand the rigours of winter; they remain dormant in the soil, or on the wheat stubble, until the spring. Then they germinate, and on the short hypha, or promycelium, that each cell produces are formed four basidiospores. These start the cycle of operations over again, though not, as we shall see in a moment, by direct infection of a wheat plant.

To examine the complex life cycle of *Puccinia graminis* in a little more detail we should start, not with the wheat plant, but with a common shrub called the barberry (*Berberis vulgaris*). This is the plant that the basidiospores of *P. graminis* attack in the spring. Released in their millions from the germinating teleutospores they are carried away by the wind and settle everywhere. If one should drop in a patch of moisture it soon germinates, a tiny hypha growing out of one end. The amount of food carried in the spore, however, is negligible, and unless the germ hypha can find nourishment it quickly dies. *Puccinia graminis* is an obligate parasite; moreover, it is strictly host specific. The only thing that can nourish the germ hypha from the basidiospore is the living protoplasm of the barberry, and nothing else will do.

The wastage of basidiospores must be colossal, for all the spores that fail to alight on a barberry bush, and they must be the vast majority, must die. This does not matter to the fungus, for there are plenty of basidiospores, and some, at least, must find their target. Those that do fall on a barberry leaf germinate successfully, and the barberry bush is infected.

Within three or four days of the barberry leaves being infected, hyphae of the fungus gather at points situated just below the epidermis, mainly on the upper sides of the leaves. At each of these points a flask-shaped structure called a pycnium is formed, containing a mass of hyphae from the ends of which spore-like bodies called pycniospores are cut off. The pycnium

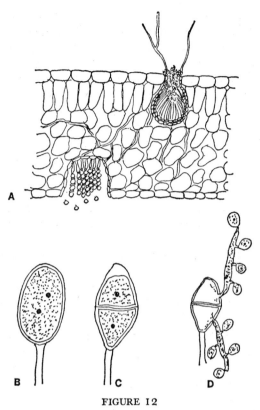

FIGURE 12

Puccinia graminis. *A, part of infected barberry leaf, showing
a pycnium with pycniospores and receptive hyphae above
and an aecidium with aecidiospores below. B, uredospore;
C, teleutospore; D, teleutospore germinating and producing
promycelia and basidiospores. Various magnifications.*

also contains a sugary fluid honey dew which oozes in droplets from an opening at the top of the pycnium (Fig. 12a).

Besides the hyphae which cut off the pycniospores the
pycnium contains some longer hyphae which stick out of the
opening at the top. These are called receptive hyphae, and
the reason for this name will become apparent later.

A little later than the pycnia, round spots begin to appear,
usually on the lower leaf surface. These are bright orange-yellow
in colour, and examination with a hand lens will generally

show that each is surrounded with a narrow white margin. These are called aecidia. They are like little inverted cups, and microscopical examination shows that they contain chains of bright orange-yellow spores, called aecidiospores. The aecidio-spores in a chain alternate with disjunctor cells whose function is made clear by their name; by their disorganisation they set free the aecidiospores, which are immediately carried away from the barberry leaf by the wind.

The basidiospores of *Puccinia graminis* are uninucleate, and when they infect the barberry they give rise to monokaryotic mycelia. The cells that produce the aecidia, and the aecidio-spores themselves, however, are dikaryotic. Sometime between the infection of the barberry plant and the formation of the aecidia the monokaryotic mycelium must be converted into a dikaryon. How this happened was for a long time a mystery, until the work of the American mycologist Craigie showed how it was done.

We now know that *Puccinia graminis* is heterothallic, so that two cells of compatible mating type must meet before the dikaryon can be established. This is where the pycniospores produced in the pycnia play their part. If a pycniospore can attach itself to a receptive hypha jutting out of a pycnium of the opposite mating strain its nucleus passes into the receptive hypha, forming a dikaryotic cell. The introduced nucleus then divides, one of its daughter nuclei passing into the adjoining cell, making that one dikaryotic. This is repeated from cell to cell until the whole mycelium becomes dikaryotic.

The only problem is the transport of a pycniospore to a compatible receptive hyphae, for those arising from the same pycnium clearly will not do. It appears that insects that feed on the honey dew produced by the pycnia are the main agents in this; picking up the pycniospores on their bodies or legs they carry them to another pycnium, where they are rubbed off on to the receptive hyphae. Rain water, and possibly even air currents, may serve the same purpose. This is not the only way in which the dikaryon can be established, for hyphal fusions between two monokaryon mycelia, of opposite mating strains, infecting the barberry leaf, can also do the job.

The aecidiospores of *Puccinia graminis* are formed in the spring, just when the wheat is ready for attack. The strange thing about them is that they are unable to infect other barberry plants, but they are highly infective for wheat. On lodging on the stem or leaf of a wheat plant they germinate, and the germ hypha that grows out of the aecidiospore finds its way into the wheat plant, entering through one of the stomata, or ventilation pores, in the epidermis. Once inside, the germ hypha gives rise to a mycelium that runs through the tissues of the wheat plant, passing between the cells, from which it draws nourishment by means of tiny suckers called haustoria which it intrudes into the cells.

The mycelium in the wheat plant gives rise, as we have seen, to the uredospores. A cushion of hyphae is formed just beneath the epidermis of the host, on which the uredospores are formed in crowded groups. The uredospores are one-celled, rusty red in colour, and each is provided with a stalk. When mature they drop off their stalks and, as we have seen, are easily disseminated through the wheat crop by the wind. The uredospores are harmless to the barberry, but they attack wheat avidly.

The mycelium that gives rise to the uredospores is strictly dikaryotic, as also was the aecidiospore from which it came. Similarly, the dark-coloured teleutospores that are presently formed are also didaryotic, at first, with a pair of nuclei in each of their two cells. This point is important if we are to understand what follows.

I said that the teleutospores were dikaryotic at first. Soon after they are formed, however, the two nuclei in each cell fuse, so that the cells become diploid.* In this diploid state the

---

*Normally the nucleus of a cell possesses a double set of chromosomes, the rod-shaped bodies which carry the hereditary information. Such nuclei are said to be diploid. When a nucleus divides by the process called mitosis, the chromosomes also divide lengthways into two, one half of each chromosome going to each daughter nucleus. This ensures that every nucleus formed by division shall have an identical set of chromosomes. The number of chromosomes per nucleus is constant for a given species. At some time before the sexual fusion of nuclei, or karyogamy, occurs there is a special division of the nucleus known as meiosis or reduction division, during which the chromosome number is halved; the nucleus is then said to be haploid. Karyogamy restores the diploid condition. In the fungi, a cell of the dikaryotic mycelium contains two *haploid* nuclei.

teleutospore passes into its winter rest. When it germinates in spring each cell gives rise to a short promycelium, and at the same time its nucleus divides into four. This is a reduction division in which the number of chromosomes is halved, so that four haploid nuclei are formed. These four nuclei pass into the promycelium, and cell walls form between them, so that the promycelium is divided into four monocaryotic cells, each with one haploid nucleus. Each cell of the promycelium then forms a single basidiospore (Fig. 12d), which in due course is shot off the promycelium and, if it is lucky, is blown by air currents to a barberry bush.

Puccinia graminis, in common with all the other rust and smut fungi, belongs to the class Teliomycetes of the sub-division Basidiomycotina. The Teliomycetes are characterised by having a compound structure, consisting of a teleutospore with its attendant promycelia, taking the place of the basidium found in the rest of the Basidiomycotina. The nearest relatives of the Teliomycetes seem to be the Phragmobasidiomycetidae, with their basidia of more than one cell, such as the jew's ear fungus described in Chapter 3.

Puccinia graminis shows the phenomenon of host specificity – the specialisation of a parasite to attack a limited range of host plants – to a higher degree than any other fungus. The species *P. graminis* attacks only cereals and grasses, all members of the family Gramineae. The specialisation, however, goes a great deal further than this, for *P. graminis* can be divided into a number of sub-species each of which is limited to one species of host; thus, *P. graminis tritici* attacks only wheat, *P. graminis hordei* is limited to barley, and so on. The various sub-species differ only minutely from one another apart from their ability to infect different host species.

Even this is not the end of the story, for within the sub-species *Puccinia graminis tritici* there are more than two hundred physiological races all differing from one another only in their ability to attack different commercial varieties of wheat. The existence of these different physiological races make the control of *P. graminis* extremely difficult.

The control of wheat rust has been a problem since the days of the Romans, who, at their annual feast to the corn god Robigus, held at the end of April when the rust was beginning to appear, sacrificed a red dog on an altar soaked in wine. This ceremony was craftily designed. The red colour of the sacrifice put Robigus in mind of the red rust, and at the same time the smell of the burning dog reminded him that he ought to chain up the dog star Sirius, which was supposed to have a malign influence on the rust resistance of the wheat. Whether or not the ceremony was effective depends, I suppose, on one's theological views, but at least it gave the Romans an excuse for a party on a grand scale, to which, if history is to be believed, they were not averse. No doubt it was enjoyed by everyone – except the dog.

The only effective way of controlling rust in wheat is by growing a rust resistant variety. The many commercial varieties of wheat differ very much in their resistance to attack by rust: some are highly susceptible while others are scarcely attacked at all. It is true that some of the varieties resistant to rust do not possess other desirable qualities such as high yield and good milling properties, but it is possible by selective breeding to combine these characteristics in one wheat. This was done by Professor Biffen of Cambridge in the celebrated series of 'Yeoman' wheats.

One might think from this that the rust problem is solved for the farmer: all he has to do is to grow a resistant variety of wheat and he can cock a snook at the rust fungus. Unfortunately, this is not so. Remember, there are over 200 physiological races of *Puccinia graminis tritici,* all differing in their virulence for different varieties of wheat. A rust-resistant wheat cannot show immunity to all of them. A variety of wheat may resist local races of rust, but *P. graminis* is cosmopolitan, and sooner or later a physiological race will turn up which can attack the 'resistant' wheat which is being grown. All the work of the plant breeders is useless, and they must start again.

There is worse to follow. Even if a variety of wheat were found that is resistant to all known physiological races of rust, there is no guarantee that it will remain so for long. New physiological races are coming into being all the time. A variety of wheat that is resistant today may be susceptible tomorrow. This is because new physiological races of *Puccinia graminis* are continually evolving from old ones, so that the plant breeders are kept constantly on the alert, with the fungus one jump ahead of them all the time.

There are two ways in which the new physiological races may arise. One is by mutation : a sudden change in the genes that determine the hereditary characteristics of an individual. Mutation we cannot prevent, for it is a natural process beyond our control. The other way is by hybridisation : the mating of two different individuals to produce a third, different from either parent.

We can do a little about this second possibility, for there is one place where hybridisation is likely to take place, and that is on the barberry bush. It is here that monokaryotic mycelia from two different basidiospores meet and between them set up the dikaryon. If we could eradicate the barberry we could at least stop this kind of variation at source.

It has been known at least since the middle ages that barberry has some connexion with rust, and for centuries farmers have been advised to destroy the barberry bush whenever they found one growing in their hedgerows. Unfortunately, eradication of the barberry is well nigh impossible, for it is a common shrub. Moreover, because of the distance to which spores of *Puccinia graminis* can be carried by the wind it is of little use to destroy the barberry on one farm if it is allowed to flourish on the next – or, for that matter, on the next but twelve !

One might think that if it were possible to exterminate the barberry completely from the surface of the earth one would get rid of the rust, but this is not so. In the United States a vigorous campaign for the extermination of the barberry has been in operation for many years. This appears to have reduced the incidence of rust appreciably in the Northern part of the U.S.A. but not in the South. This is because in the southern states the winters are mild enough to allow the uredospores of *Puccinia graminis* to survive and infect the following year's crops, while the more rigorous winters in the North will not permit this. Consequently, in the South the rust can miss out its aecidiospore, teleutospore, and basidiospore stages completely and survive by forming uredospores alone. Since these are infective only for wheat, the barberry is not needed.

When we come to consider the other species of *Puccinia*, and the rust fungi as a whole, we find that it is quite a common thing for the fungus to miss one or more of its spore stages; some of them do it as a normal routine. The hollyhock rust (*Puccinia malvacearum*) exists in its teleutospore stage only: there are no aecidiospores or uredospores. The same is true of *P. adoxae*, a rust attacking that rather delightful little plant the moschatel or town hall clock (*Adoxa moschatellina*).

In *Puccinia graminis* we have a rust that normally produces its full panoply of spore forms: aecidiospores, uredospores, teleutospores, and basidiospores. Such a rust is called a macrocyclic rust, a term applied to any rust which form one or more *binucleate* (dikaryotic) spore types in addition to the teleutospores. Rusts in which the teleutospores are the only binucleate spores are classed as microcylic rusts.

Puccinia graminis, in common with many other rust fungi, has two separate hosts in its life cycle, and is therefore called a heteroecious rust. Some rust fungi, such as *Puccinia malvacearum*, have one host only; they are autoecious. It might seem logical to suppose that, in a heteroecious rust, the two host plants would be related to one another fairly closely, but this is usually not so. There is no relationship between the barberry and the wheat, except that both are flowering plants. Sometimes the two hosts are ever farther apart. *Cronartium ribicola* produces its aecidiospores on the white pine (*Pinus strobus*), a conifer, and its uredospore and teleutospore stages on currants and gooseberries.

Where a rust has two hosts the host on which the teleutospore stage occurs is called the principal host.

The rust fungi form the class Uredinales. They are divided into various genera according to the form of the teleutospores. Thus, in *Puccinia* the teleutospores are two-celled, in *Uromyces* they have one cell each, in *Phragmidium* they have several cells arranged in a line, and so on. Rust fungi that lack the teleutospore stage have to be classified on their aecidiospores or uredospores into 'form genera', the word 'form genus' indicating that the classification is only a matter of convenience, without any implications regarding relationship.

It is an interesting fact that, although *Puccinia graminis* is such a devastating parasite of wheat, it seldom kills its host plant. This, when you come to think of it, is sensible behaviour. Since *P. graminis* is strictly an obligate parasite (it *has* been grown now in Australia away from its host, with great difficulty, but that is splitting hairs) it follows that if the host dies the fungus must die too. It cannot survive the death of its host plant, even by living on its rotting remains. It therefore behoves it not to kill the goose that lays the golden eggs.

The parasitic relationship between *Puccinia graminis* and its host is extremely delicately balanced. The fungus extracts as much as it can from the host plant, so that the health of the host is badly affected, as is the yield of grain. The fungus interferes with the intake of water by the host, and also takes nutriments which the host needs for itself. The net result is that the wheat plant is stunted and produces little or no corn. Only in rare cases, however, is it actually killed: the fungus is too cunning for that.

One often finds this kind of delicate relationship between a strictly parasitic fungus and its host. On the other hand, fungi which are less specialised for the parasitic life often kill their host plants out of hand. *Pythium* is an example. The sixty-odd species of *Pythium,* as far as we know, normally live as saprobes in the soil, but many of them can be destructive parasites of seedlings if occasion offers. As parasites they have no finesse: they kill off their host seedling as quickly as may be and then gorge themselves on its dead and rotting remains. When it has been squandered they return to a saprobic life in the soil until another suitable host turns up. The difference in technique between *Pythium* and *Puccinia* is the difference between an amateur and a professional.

THE SMUTS AND BUNTS

The smut and bunt fungi (order Ustilaginales) are related to the rusts, which they resemble in many ways, though there are also important differences. There is no essential difference between a smut and a bunt, and both names may cover diseases produced by the same fungus, as in bunt or stinking smut of wheat, caused by species of *Tilletia*. There are three families of smut and bunt fungi, two of which, the Ustilaginaceae and the Tilletiaceae, are well known and widely distributed, while the third, the Graphiolaceae, is little known and has a restricted distribution.

The rusts are remarkable for having, typically, several different spore forms. These are not found in the Ustilaginales, in which the typical spore form is the teleutospore with its attendant basidiospores, though conidia may also be produced.

The Ustilaginales are all parasites on plants, and the diseases they cause can be of serious economic importance. Stinking smut of wheat, caused by *Tilletia caries* and *T. foetida*, rivals the black stem rust of wheat as a world agricultural menace. Some of the Ustilaginales are extremely common and widespread; among these may be counted corn smut (*Ustilago maydis*) attacking maize, loose smut of oats (*U. avenae*), and onion smut (*Urocystis cepulae*).

Unlike the rusts, the Ustilaginales are not obligate parasites. Many of the smut fungi have been successfully grown in laboratory cultures on artificial culture media.

The basidiospores of the Ustilaginales have only one nucleus. In some of the smuts they may germinate to give rise to a monocaryotic mycelium, but this is always short-lived, the dikaryotic condition being quickly established by fusion between hyphae. In the family Tilletiacaea particularly the only monocaryotic cell may be the basidiospore itself, the basidiospores conjugating in pairs to produce binucleate cells from which a dikaryotic mycelium is produced immediately.

The mycelium of the smut fungi does not develop so profusely as that of some parasitic fungi, but it does spread through the tissues of the host plant, passing mainly between the cells of the host, though in some species, such as the corn smut, the hyphae may enter the host cells here and there. As the mycelium develops conidia may be formed on hyphae near the exterior of the host. Smut fungi may also reproduce freely by a process called budding. Either a conidium or a basidiospore may reproduce in this way, a small protrusion growing out of the side of the spore and increasing in size until it is large enough to break away and assume an independent existence.

When the time comes for the formation of teleutospores the hyphae of a smut fungus usually form many short cells which become surrounded with thick, spherical walls; these are the teleutospores. They are usually dark brown to black in colour, which is why these fungi are called smuts; the host plants often have a dirty, sooty appearance when the dark masses of teleutospores are forming. As in the rusts, the teleutospores of the smuts are binucleate at first, but the two nuclei eventually fuse to form a diploid nucleus.

The germination of the teleutospores is favoured by low temperatures; this is unusual, for the germination of most fungus spores is promoted by warmth. On germination the diploid nucleus of the teleutospore undergoes meiosis, forming four haploid nuclei. A short hypha grows out of the spore; this is called the promycelium as in the rusts. The way in which the basidiospores are produced on the promycelium depends on the family to which the smut fungus belongs. In the Ustilaginaceae the promycelium is divided into four cells. The nucleus of each cell divides, and one daughter nucleus in each cell passes into a

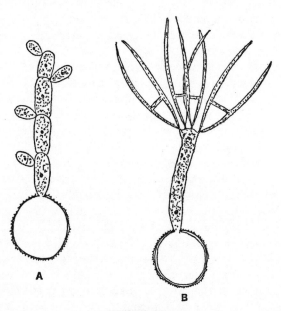

A

B

FIGURE 13

Teleutospores of the Ustilaginales. A, Ustilago, *showing four-celled promycelium with basidiospores. B,* Tilletia, *showing pairs of sickle-shaped basidiospores conjugating. Greatly magnified.*

bud that forms from the side of the cell; this bud becomes the basidiospore (Fig. 13a). The nuclei that are left behind in the cells of the promycelium may divide again, so that more basidiospores may be produced; in this point the smuts differ from the rusts, in which the cells of the promycelium are left empty after the formation of the basidiospores, so that a further crop is impossible.

In the Tilletiaceae the behaviour during basidiospore formation is somewhat different. The promycelium is not divided into separate cells. The basidiospores are formed at the end of the promycelium, and there are usually eight of them, although their number varies. The basidiospores are usually long and narrow. Copulation tubes are formed between pairs of basidiospores, the nucleus of one of them passing into the other to form a binucleate spore (Fig. 13b) on which crescent shaped, binucleate conidia are formed; the conidia then infect the appropriate host plant, a dikaryotic mycelium being formed. Instead of forming conidia a binucleate basidiospore may infect the host plant directly, again forming a dikaryotic mycelium.

7 The cup fungi

The cup fungi belong to the sub-division Ascomycotina, another of the major groups into which the fungi are classified, equal in status to the Basidiomycotina with which we have up to now been concerned. In the Ascomycotina the spores, which are called ascospores, are formed inside special cells called asci, and each ascus usually contains eight ascospores. Their formation is preceded by a rather complex sexual process which I will describe in a moment. In the Ascomycotina the asci are nearly always contained in a fruit body or ascocarp which, in the cup fungi, usually takes the form of a small cup or apothecium with the asci covering the inner surface. The size of the cup varies; in the orange peel fungus (*Aleuria aurantia*), it may measure up to 10 centimetres (4 inches) in diameter, while in the smaller cup fungi the cup may be almost microscopic.

In some cup fungi, such as the morel (*Morchella esculenta*), the apothecium is not cup-shaped. In the morel it forms a wrinkled cap on the end of a stalk, while in the earth tongues (*Geoglossum*) it is club-shaped.

The cup fungi are collectively placed in the class Discomycetes. Most of them are saprobes growing on decaying wood or other plant material, on dung, or on the surface of the soil, but some, such as *Monilinia fructigena,* the cause of brown rot of apples and other fruit, are destructive parasites.

THE LIFE HISTORY OF A CUP FUNGUS

The sex life of the Ascomycotina is extremely complex, and the full story has only been worked out in detail for a few species out of the many thousands that have been described. One of the few is *Pyronema omphalodes*, formerly known as *P. confluens*, a little cup fungus that shows a distinct liking for soil that has been heated; its pink fruit bodies are often found on the surface of the soil where a bonfire has burned, and it also occurs in pots of soil that have been sterilised by heat. *P. omphalodes* has been extensively used for research on the Ascomycotina for more than half a century, so it is not surprising that the details are fairly well known for this particular species; this makes it suitable as an introduction to the cup fungi as a group.

The mycelium of *Pyronema omphalodes* consists of a delicate weft of septate hyphae (the Ascomycotina, like the Basidiomycotina, always have hyphae which are divided into cells by cross-partitions or septa). The mycelium occurs on burnt ground, or on fallen leaves. *P. omphalodes* is a saprobe, living on decaying organic matter in the substratum on which it is found.

Unlike most of the Ascomycotina, *Pyronema omphalodes* does not have any asexual spores. This is unusual, for in the Ascomycotina as a group we find that asexual spores in the form of conidia (spores produced exogenously on the end of a branch and not enclosed in a sporangium) are usually produced in great profusion, particularly in the summer when the mycelium is growing strongly. Asexual spores are essentially an accessory means of reproduction, designed to propagate and disseminate the fungus in the spring and summer, when conditions for growth are good.

The male and female sex organs of *Pyronema omphalodes* are called antheridia and ascogonia respectively. They are formed on separate hyphae, and both contain many nuclei. The antheridia are one-celled and columnar, or somewhat club-shaped. The ascogonia each consist, when mature, of two cells. The lower cell, which is somewhat bulbous, is called the oögonium, and on top of it there is a slender, elongated cell called the trichogyne. Antheridia and ascogonia are usually formed close together, so that they stand side by side.

P. omphalodes, like many of the Ascomycotina, shows the phenomenon of heterothallism (see Chapter 1). There are two compatible strains, the plus and minus strains, and the ascogonium from a plus strain can only be fertilised by an antheridium of a minus strain, or *vice versa.*

The trichogyne of the ascogonium grows towards the antheridium and makes contact with its tip (Fig. 14a). At the point of contact between them the walls of the antheridium and trichogyne dissolve, so that there is an open connexion between the two organs. Nuclei from the antheridium pass into the trichogyne and, travelling down it, enter the oögonium. To do this they must pass through the cell wall separating trichogyne and oögonium; this is a simple matter, as the septa between the cells of fungi have perforations which will allow nuclei to slip through.

The passage of the contents of the antheridium into the oögonium via the trichogyne constitutes plasmogamy, the first stage in sexual reproduction. One might expect this to be followed at once by karyogamy, the fusion of male and female nuclei, but this does not happen. According to classical research the male and female nuclei pair off with one another, occupying positions side by side, but recent work has thrown some doubt on this. At any rate, there are no fusions of nuclei, and the ascogonium at this stage contains both plus and minus nuclei, which tend to arrange themselves around the periphery of the ascogonium.

The next stage is the outgrowth of numerous small papillae from the wall of the ascogonium, each papilla commonly arising opposite a group of nuclei (Fig. 14b). Nuclei begin to pass into the papillae, which elongate until finally they form a mass of hyphae arising from the ascogonium. These are called ascogenous hyphae. Meanwhile, the nuclei in the ascogenous hyphae, and also those left in the ascogonium, undergo simultaneous divisions. The nuclei in the ascogenous hyphae are now arranged in pairs, each pair consisting of one plus and one minus nucleus. At first the ascogenous hyphae are nonseptate, but presently septa form, cutting the ascogenous hyphae into cells in such a way that each cell contains a plus and a minus nucleus.

The cell at the end of each ascogenous hypha now bends over, forming a hook, or crozier as it is sometimes called from its resemblance to the crozier carried by a bishop on ceremonial occasions. The two nuclei in the crozier each divide once, so that the crozier now contains four nuclei. Septa are then formed dividing the crozier into three cells: a downwardly pointing terminal cell with one nucleus, a penultimate cell, occupying the bend in the crozier, with two nuclei, and a third

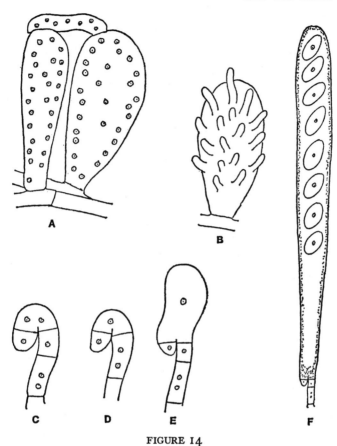

FIGURE 14

Diagram showing how the ascogenous hyphae and asci are formed in Pyronema omphalodes. *A, Ascogonium and antheridium, the trichogune of the ascogonium arching over and making contact with the top of the antheridium; B, Ascogenous hyphae beginning to grow out of the fertilised oögonium; C, a crozier formed at the tip of an ascogenous hyphae; D, fusion of the two nuclei in the penultimate cell of the crozier; E, growth of the young ascus; F, a mature ascus. Varying magnifications.*

cell with one nucleus, often called the stalk cell (Fig. 14c). The binucleate penultimate cell is the ascus mother cell – the cell which gives rise to the ascus.

If you will refer to Fig. 14 for a moment, you will see that, owing to the peculiar arrangement of the plus and minus nuclei in the ascogenous hypha, the penultimate cell contains a nucleus of each mating type. Furthermore, the nuclei of the terminal cell and the stalk cell contain nuclei of opposite mating types. These facts are important in view of what happens next.

The two nuclei in the penultimate cell now fuse; this is the act of karyogamy for which we have been waiting, and its occurrence makes the sexual reproduction complete (Fig. 14d). The penultimate cell begins to enlarge (Fig. 14e); it becomes, in fact, the young ascus. It becomes wider and very much longer, for in most of the cup fungi the asci are long and narrow. While it is elongating the single nucleus produced by the fusion of the plus and minus nuclei in the penultimate cell begins to divide; altogether there are three divisions, so that eight nuclei are formed. The first two of these divisions are meiotic and the final division is mitotic, so that the developing ascus now contains eight haploid nuclei.

Each of the ascogenous hyphae has now produced an ascus at its tip. The terminal cell may fuse with the stalk cell to form a cell from which another crozier is developed, and the process may again be repeated, so that one ascogenous hypha may give rise to several asci.

As the ascus elongates its cytoplasm collects round the eight nuclei, forming eight ascospores; a certain amount of cytoplasm (the epiplasm) is left over, and as the ascus enlarges to its final size this is spread out in a thin layer just inside the ascus wall, the centre of the ascus being occupied by a large fluid-filled vacuole in which the ascospores are suspended in the upper half of the ascus (Fig. 14f). At first the epiplasm contains a substance called glycogen, but later this disappears, probably being converted to a sugar called glucose. Owing to a phenomenon known as osmosis, the presence of glucose in the ascus causes water to enter it, so that it becomes notably distended. The pressure of the water leads to the bursting of the tip of the ascus. In some Ascomycotina the tip of the ascus may tear, but in Pyronema and its relatives the ascus opens by means of a little cap, or operculum, releasing the ascospores which are squeezed out of the ascus with the escaping water, and flung to a distance of two or three centimetres. The germination of the ascospores starts the life history over again.

As yet I have said nothing about the development of the cup-shaped fruit body that contains the asci. The ascogenous hyphae springing from the ascogonium grow upwards in a close bunch. As they do so, branches from the mycelium just below the ascogonium grow up round the bunch of ascogenous hyphae, forming a closely interwoven sheath in the form of a cup. There are two points to note about this. The first is that the cup is not formed from the fertilised ascogonium but from the vegetative mycelium below it, and the second is that all the ascogenous hyphae enclosed in any one cup spring from one ascogonium.

The fertile layer, or hymenium, of asci forms a lining to the inside of the cup, the asci standing up vertically, mixed with sterile hairs called paraphyses. When the ascospores are discharged they are shot straight up into the air, the wind normally carrying them clear of the cup before they fall. In some of the larger cup fungi it is possible to see clouds of ascospores being given off in puffs as groups of asci discharge their spores together.

The asci themselves, before they discharge their ascospores, became positively phototropic (grow towards the light). This ensures that the ascospores will be shot in the direction of an open space, a point that is essential for their dispersal.

We are now in a position to compare the life history of a cup fungus with that of a typical member of the Basidiomycotina, as, for instance, the toadstool *Coprinus*. You will remember that the haploid basidiospores of *Coprinus* give rise to a monokaryotic mycelium, and that the monokaryophase is short lived, becoming converted to a dikaryon by fusion with another monokaryotic mycelium, or with an oidium, before the formation of the fruit body. In other words, for the greater part of its life cycle *Coprinus* is dikaryotic. Plasmogamy occurs when the monokaryon is converted to a dikaryon, but karyogamy is postponed right until the end of the life history when the basidiospores are about to be formed.

In a cup fungus such as *Pyronema omphalodes* we find the same interval between plasmogamy and karyogamy, but it is much shorter. The germination of the ascospore gives rise to a monokaryotic mycelium which occupies the greater part of the life history of the fungus; in most Ascomycotina, though not as it happens in *P. omphalodes*, the monokaryon even reproduces itself extensively by conidia. Towards the end of the life cycle the sex organs are formed, and plasmogamy occurs when the ascogonium is fertilised by the antheridium. Then comes the dikaryophase in the form of the ascogenous hyphae, which is brought to an end with karyogamy in the young ascus. The dikaryophase in the Ascomycotina is only a small, though none the less important, part of the life history.

These points may seem to be of little importance, but they become significant when we consider the evolution of the fungi. There is little doubt in the minds of most mycologists that the Basidiomycotina were derived from the Ascomycotina a long time ago; possibly from one of the more primitive Ascomycotina such as *Taphrina*. An ascus is really very like a basidium, believe it or not! As ascus may be defined as 'a cell in which nuclear fusion is followed by meiosis and endogenous spore formation', while the corresponding definition of a basidium needs the alteration of only one word: exogenous for endogenous.

OTHER CUP FUNGI

The cup fungi are a fascinating group, worthy of more attention from amateur collectors. Many species are quite common, producing their fruit bodies mainly in the autumn, and some of them are edible. Among the edible species I have already mentioned the orange peel fungus, *Aleuria aurantia* (Fig. 15a), which is to be found growing on the ground in woods in autumn. Its cup is up to 10 centimetres (4 inches) in diameter, and of a bright orange colour; it looks from a distance very like a piece of orange peel casually thrown on the ground.

The brown elf cup, *Peziza vesiculosa*, may be found from spring until late autumn on rich soil in gardens and elsewhere. Its fruit body is somewhat irregular in shape, brown in colour, and about 5 centimetres (2 inches) in diameter; the outside of the cup is pale coloured and somewhat mealy. The fruit bodies tend to grow in groups. *P. badia* is similar in appearance but rather larger, with cups 7.5 to 10 centimetres (3 to 4 inches) in diameter;

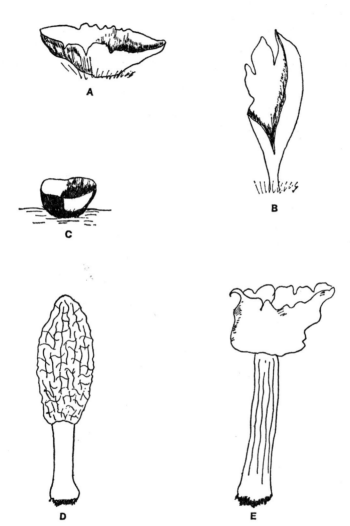

FIGURE 15

Some cup fungi. A, the orange peel fungus (Aleuria
aurantia); *B, the hare's ear* (Otidia leporina); *C, the black
bulgar* (Bulgaria inquinans); *D, the common morel*
(Morchella esculenta); *E,* Helvella crispa, *one of the false
morels. Not drawn to the same scale.*

it can readily be distinguished from *P. vesiculosa* because its cups are of a uniform brown colour inside and out. The scarlet elf cup, *Sarcoscypha coccinea,* has a cup with a bright scarlet interior, about 2.5 centimetres (1 inch) in diameter, and provided with a stalk. The cups may be found in winter on dead twigs.

The genus *Otidea* is characterised by having a cup shaped like the ear of a rabbit. The hare's ear, *Otidea leporina* (Fig. 15b), has a cup about 5 centimetres (2 inches) high and grows among spruce needles. The cups are usually found in clusters, and are brown inside and yellowish on the outside.

The black bulgar (*Bulgaria inquinans*) is a common cup fungus that grows on oak, elm, and beech logs. Its fruit body is jet black and about 2.5 centimetres (1 inch) in diameter, with a distinctive rubbery consistency (Fig. 15c).

In the earth tongues (Geoglossaceae) the fruit body is club-shaped and flattened, with the hymenium of asci covering the surface. *Geoglossum cookeianum,* for instance, has an olive-green fruit body from 2.5 to 5 centimetres (1 to 2 inches) high; it grows on the surface of the soil in woods and copses.

THE MORELS

The morels (*Morchella*) are a genus of edible fungi that do not look as if they were related to the cup fungi at all. The common morel, *Morchella esculenta* (Fig. 15d), has a yellowish brown, egg-shaped cap, from 2.5 to 5 centimetres (1 to 2 inches) long, on top of a yellowish stalk from 5 to 10 centimetres (2 to 4 inches) high. The cap has a much wrinkled surface bearing the hymenium of asci, while the stalk is smooth and brittle. The fruit bodies appear in spring, usually growing where the soil is rich along hedgerows and on wayside banks. They are excellent to eat, and can be dried and stored for winter use. The mycelium of the common morel can be grown in fermentation tanks and its use for making canned 'mushroom' soup is being developed commercially.

The peculiar fruit body of *Morchella* is regarded as a stalked cup which has turned inside out, so that the hymenium is fully exposed.

THE FALSE MORELS

The false morels belong to the same family (Helvellaceae) as the true morels, which they resemble in their general features. Like the morels, they have fruit bodies consisting of a cap borne on a stalk, but the cap is variable in form; in some species it may consist of a cup supported on a stalk, but usually it is irregularly lobed, presenting a somewhat untidy appearance. The stalk is often grooved or wrinkled rather than smooth. The false morels usually produce their fruit bodies in the summer and autumn rather than in the spring. Although the false morels are said to be edible they should be eaten with considerable caution, for some people are upset by them. In any case they must never be eaten raw since they contain a poisonous substance that is destroyed by cooking. My advice is to avoid them, for there are plenty of good edible fungi without taking a risk on a false morel.

The commonest British false morel is *Helvella crispa* (Fig. 15e), the fruit bodies of which occur in both spring and autumn, usually in the grass alongside paths in woods. It has a saddleshaped cap, much lobed, and pale brown to cream in colour. It is from 7.5 to 10 centimetres (3 to 4 inches) high, with a deeply furrowed stalk.

THE TRUFFLES

The truffles (Tuberaceae) are a group of hypogeous (underground) fungi that are much prized as delicacies by mycophagists. The fruit bodies are more or less spherical, and are produced entirely underground, at a depth of from 2.5 to 15 centimetres (1 to 6 inches). They measure from 2.5 to 12.5 centimetres (1 to 5 inches) in diameter, according to the particular species, and are found buried beneath the soil in calcareous woods, particularly near oak trees.

The asci of the truffles are formed inside the fruit body. Since this is underground there is no question of the spores being dispersed by wind; they have to await the services of an animal which eats the fruit body and in due course voids the ascospores with its faeces. The animal may be man, a dog, a wild pig, or a rodent. To encourage the spore dispersal mechanism a truffle is provided by nature with the most delectable, irresistible, mouth-watering smell that ever tantalised the nose of man or animal. Pigs, in particular, love it; they can smell a buried truffle from a distance of many metres and make unerringly for the spot where it is buried.

The love of pigs for truffles is made use of in France, where trained pigs are used for truffle hunting. The pig is carried to the woods where truffles are found, or carted in a wheelbarrow, so that she will not get tired before starting work. A rope is then tied round her neck, and the hunt is on. When a truffle is located the pig starts to dig it up; she is then tied to a handy bush while her master finishes the job she has started, the pig being rewarded with an acorn or some such inferior delicacy. Pigs are specially trained for truffle hunting in Normandy and elsewhere, and a good truffling pig can fetch a small fortune.

If you have the ambition to eat a truffle – and what mycophagist has not? – and do not happen to have a pig, you need not despair. Much can be done with a rake, provided you have plenty of energy. A calcareous oakwood is the best place to try, but a beechwood on chalk or limestone will do. Insects appear to like truffles as much as mammals do, and the presence of a ripe truffle underground can often be detected by swarms of small flies hovering about a foot above the spot where it is buried.

The summer truffle (*Tuber aestivum*) is found in calcareous beechwoods; it is dark brown to blackish, with warts on its skin, and from 2.5 to 10 centimetres (1 to 4 inches) in diameter. *Tuber rufrum* is smaller, usually no more than 2.5 centimetres (1 inch) in diameter; it occurs in woodlands of various kinds, both broadleaved and coniferous. The largest British truffle is the white truffle, *Choiromyces meandriformis*, which may measure 12.5 centimetres (5 inches) in diameter. Its colour varies from cream to pale brown, and it is found in oakwoods.

8 The flask fungi

In the cup fungi (class Discomycetes) the fruit body is usually a cup containing the asci though, as we saw in the last chapter, it can have other shapes. In the flask fungi (class Pyrenomycetes) the asci are enclosed in a flask-shaped structure with an opening called the ostiole at the top. The flask is called a perithecium. The cup, or apothecium of a cup fungus can be quite large, reaching a diameter of 10 centimetres (4 inches) in the orange peel fungus, but the perithecia of the flask fungi are always small; usually a hand lens is needed to see them, though they are often aggregated together to form larger structures.

The flask fungi, like the cup fungi, belong to the Ascomycotina, and the general story of the fertilisation of the ascogonium by the antheridium and the formation of ascogenous hyphae and asci follows the same course as was described in the last chapter, though there are many differences in detail. These differences, which are also met with in the cup fungi, need not concern us here.

In the simpler flask fungi the perithecia are produced singly on the mycelium, but in many genera the perithecia are gathered together in a structure called a stroma (plural, stromata). A stroma is a dense structure, formed of closely interwoven hyphae, in which the perithecia are embedded; sometimes the perithecia, instead of being sunken, are scattered on the surface of the stroma. A stroma may reach a considerable size; in the insect parasite *Cordyceps norvegica* the club-shaped stromata are up to 20 centimetres (8 inches) long.

THE SPHAERIALES

The Sphaeriales are the largest order of the flask fungi, containing several thousand species. Many of them are saprobes growing on rotting vegetation, on the soil, or on the dung of animals, but the order also includes some important plant parasites. The Sphaeriales are characterised by having their perithecia, and their stromata if any, of a leathery or woody consistency, and of a dark brown or black colour.

One of the simplest and commonest of the Sphaeriales is *Sordaria fimicola,* a saprobe which grows on the dung of herbivorous animals. It is a regular constituent of the dung flora, and it is also easily grown in the laboratory on agar culture media (agar is a substance, derived from seaweed, that is added to culture media in order to make them set to a jelly).

The mycelium of *S. fimicola,* as in all the Ascomycotina, consists of septate hyphae. The flask-shaped perithecia, which are about half a millimetre high, are formed singly; there are no stromata (Plate 8). It is a strange fact that, in spite of the very large amount of work that has been done on *Sordaria* as a convenient and easily obtained fungus for laboratory research, nobody knows very much about the details of its sexual reproduction. One worker has reported observing contact between antheridia and ascogonia, while others have shown that fusion between vegetative hyphae occurs, nuclei passing from one hypha to the other, without the formation of organised sex organs at all. When qualified experts disagree violently about the subject of their observations the true answer is usually six of one and half a dozen of the other; probably, though we cannot say for certain, *S. fimicola* reproduces sexually in both ways.

S. fimicola, like *Pyronema omphalodes,* is somewhat peculiar in one respect: it does not reproduce asexually by means of conidia at all.

The interior of the perithecium of *S. fimicola* is occupied by asci in all stages of development. As each ascus becomes ready to discharge its ascospores it elongates rapidly, its tip protruding through the ostiole of the perithecium. The tip of the ascus bursts open, and the eight ascospores are ejected to a distance of as much as 10 centimetres (4 inches). This may not sound far, but remember that the ascospores are extremely small. When an ascus has discharged its spores it collapses and soon disintegrates, leaving room for the next ascus to push through the ostiole. The asci thus discharge their spores in regular succession. A point to notice about the spores of *S. fimicola* is that they are stuck together with mucilage so that they are discharged in a single bunch. The extra weight of the bunch compared with the weight of a single ascospore gives them greater momentum when they are discharged, so that they will travel to a greater distance against the resistance of the air.

You will remember that the asci of *Pyronema omphalodes* grew towards the light, to ensure that the ascospores were discharged into the open. In *Sordaria fimicola* the top end perithecium grows towards the light, for the same reason.

Another common member of the Sphaeriales is the pink bread mould, *Neurospora sitophila*, a fungus which sometimes causes a great deal of nuisance when it becomes rampant in bakeries. The mycelium of *N. sitophila* consists of septate hyphae which produce an abundance of conidia in branched chains; these conidia are multinucleate and are called macroconidia to distinguish them from other, smaller conidia, called microconidia, which are produced, more sparingly, in unbranched chains. *N. sitophila* can reproduce itself abundantly by means of its macroconidia; in fact, until the sexual stage was discovered in 1927, the fungus was only known in its asexual stage, to which the name *Monilia sitophila* was given.

The mycelium of *N. sitophila* is coloured from pink to red, the amount of pigment produced varying with the kind of substratum the fungus is growing on. The readiness with which it produces its macroconidia makes it a difficult mould to eradicate once a room is thoroughly infected. It can cause havoc if it gets loose in a mycological laboratory where pure cultures of fungi are kept, contaminating the cultures as fast as they are made; usually the laboratory has to be completely disinfected before the trouble is cured. Similarly, *N. sitophila* can do a great deal of damage to bread in bakeries, which are usually more difficult than laboratories to disinfect.

N. sitophila differs sharply from *Sordaria fimicola* in the abundance with which it forms conidia. It also differs from *S. fimicola* in its sexual reproduction, for *Neurospora sitophila* is heterothallic, so that no ascocarps can be produced unless mycelia of plus and minus strains meet, in spite of the fact that ascogonia develop on both strains. The ascogonium is long and coiled; it consists of an oögonial region of several cells, and a long, slender, multicellular trichogyne. There is no antheridium such as we saw in *Pyromena omphalodes*. Fusion occurs between the trichogyne and a microconidium which functions as a male gamete, giving up its nucleus to the ascogonium and thus initiating the dikaryophase. Instead of fusing with a microconidium, the trichogyne may be fertilised by a marcroconidium, or a vegetative hypha of the opposite mating strain may fuse with the trichogyne and supply the 'male' nuclei.

It may thus be seen that *Neurospora sitophila* is distinctly lax in its observance of the minutiae of sexual reproduction, and that the same end may be met in different ways according to circumstances. This casualness about sex is not at all uncommon in the Ascomycotina, which show a strong tendency towards sexual degeneracy, especially as regards the development of the male sex organ or antheridium. In some of the Ascomycotina no male gametes of any kind are produced, and the pairs of nuclei that ultimately fuse in the young ascus are pairs of ascogonial nuclei, a state of affairs called pseudogamy.

With the fertilisation of the ascogonium, development of a flask-shaped perithecium containing asci proceeds as in *Sordaria*.

Two other common species of *Neurospora* are *N. crassa* and *N. tetrasperma;* the latter, as its name implies, has only four ascospores in each ascus. All three species, and especially *N. crassa*, have been extensively used in research into the biochemistry and genetics of the fungi.

The stag's horn fungus, or candle snuff fungus, *Xylosphaera* (*Xylaria*) *hypoxylon,* is a common example of a member of the Sphaeriales in which the perithecia are developed in a stroma. This little fungus can be found at any time of the year growing on stumps of hardwood trees, or on fallen branches and twigs. It produces hard, black stromata between 2.5 and 5 centimetres (1 and 2 inches) high; they are branched like the antlers of a stag, and they have white tips when young, but as the stromata become older the tips darken (Fig. 16).

The perithecia of Xylosphaera are buried in the stroma, with just their tips projecting (Plate 9). *Xylosphaera*, like most of the Xylariaceae, is a saprobe, but a few of the Xylariaceae are parasites; *Nummularia discreta*, for instance, causes the disease known in America as nail head of apple trees.

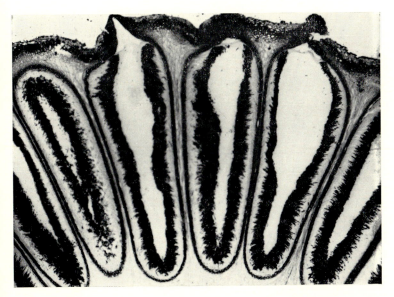

Plate 9. Photomicrograph of a vertical section through part of the tip of a mature stroma of the stag's horn fungus (*Xylosphaera hypoxylon*), showing the embedded perithecia. Magnified

Plate 10. Photomicrograph of a vertical section of the stroma of the ergot fungus (*Claviceps purpurea*), showing the embedded perithecia. Magnified

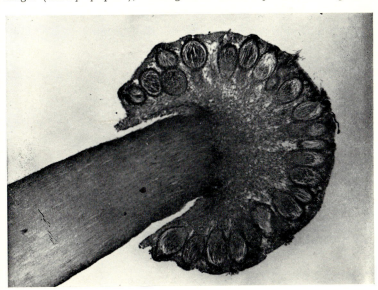

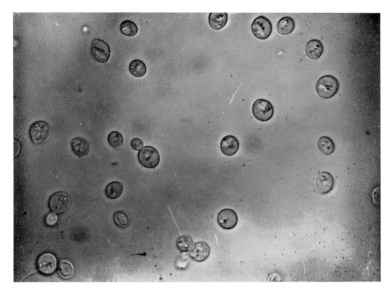

Plate 11. Photomicrograph of the bakers' yeast (*Saccharomyces cerevisiae*). Several of the cells show budding. Greatly magnified

Plate 12. Photomicrograph of a zygospore of *Absidia glauca*. The curled appendages are typical of this genus. Magnified

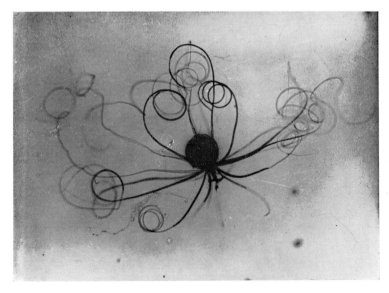

FIGURE 16

A stroma of the stag's horn fungus (Xylosphaera hypoxylon).

THE HYPOCREALES

The Hypocreales are a large order of fungi in which we place species with bright-coloured perithecia with soft or waxy walls, or bright-coloured stromata which are more or less soft in texture. They include both saprobic and parasitic species.

The coral spot fungus, *Nectria cinnabarina*, is an extremely common member of the Hypocreales. It is weakly parasitic on a number of different species of trees in Europe and America, and its fruit bodies can often be seen on dead branches of hardwood trees, as well as on old pea sticks and damp twigs. It is especially common on the wood of the sycamore.

The asexual reproductive phase of *Nectria cinnabarina* begins with the production of small, pink stromata just below the surface of the host tissue; as they develop these stromata burst through the surface of the bark, where they stand out as bright orange-pink cushions about 3 millimetres ($\frac{1}{8}$ inch) in diameter, carried on very short stalks. The surface of the stroma consists of a mass of conidiophores which bear small, elongated-

D

oval conidia. A stroma bearing masses of conidiophores in this way is called a sporodochium. The conidia, which are produced in vast numbers, reproduce the fungus during the summer months.

As summer turns to autumn, the colour of the stromata changes from orange-pink to dark red. This is because dark red perithecia have formed, covering the surface of the stroma and replacing the conidiophores. The asci within the perithecia contain two-celled ascospores.

Nectria cinnabarina is so weakly parasitic that it probably does little harm to its host. A more serious parasite is *N. galligena*, which causes a canker of apple trees.

Gibberella fujikuroi is a parasitic member of the Hypocreales which causes the 'foolish seedling' disease of rice in Eastern countries, as well as the disease known as corn red ear-rot in America. *G. fujikuroi* is the source of gibberellic acid, a substance which produces a remarkable effect on the growth of higher plants. Under its effect plant stems increase greatly in length; its discovery was due to the fact that rice plants attacked by the fungus were abnormally tall. It has since been shown that gibberellic acid is one of the growth-promoting substances that occur in higher plants.

THE CLAVICIPITALES

The Clavicipitales, a small but important order, used to be part of the Hypocreales. The order contains only one family, the Clavicipitaceae, with long, cylindrical asci each of which has a thick cap with a pore allowing the escape of the ascospores. The ascospores are long and thread-like, and in some species they break into separate portions when they have been released from the ascus, each portion germinating like a spore. The Clavicipitaceae include the important genus *Claviceps*.

ERGOT

The ergot fungus, *Claviceps purpurea*, is a widespread parasite of cereals and some grasses, and is especially common on rye. In countries where black bread, made from rye flour, is eaten extensively it may give rise to epidemics of the disease called ergotism.

The ascospores of *C. purpurea* are discharged from the asci in Spring, at the time when susceptible grasses are coming into flower. They attack the flowers of the host, causing an infection of the ovary, the developing mycelium destroying the ovarian tissues and replacing them with a mycelial mat. The ovaries appear much wrinkled and shrunken, and are known as pseudomorphs. The mycelial mat bears at its surface large numbers of conidiophores, which produce at their tips an abundance of small conidia which spread the disease to other susceptible plants. Quantities of a sweet, sticky liquid called honey dew are produced with the conidia, and this attracts insects which assist in dispersing the conidia.

Later in the season the mycelial mat in the ovary of the host becomes harder, and develops into a pinkish or purple, cigar shaped sclerotium which, being somewhat longer than the grain that bears it, sticks out of the ear of the rye or other infected grass (Fig. 17a, b). To the fanciful the grains of rye appear to be smoking cigars. The rye is then harvested, during which operation some of the sclerotia are knocked out of the ears of corn on to the ground, the remainder being harvested with the corn.

The sclerotia survive the winter in a dormant state, and in spring they germinate, each one producing from one to several purple stromata. The stromata are mushroom-shaped and about 10 mm. (⅜ inch) tall (Fig. 17c). Cavities appear just below the surface of the stromatal head, in each of which an ascogonium appears, with several antheridia arising near its base. Fertilisation and the production of ascogenous hyphae follow, and finally perithecia are formed, opening at the surface of the stroma. Each perithecium contains eight thread-like ascospores (Plate 10).

The sclerotia of *C. purpurea* are called ergots; you will find them listed under that name in pharmacological reference works, for the ergots contain a number of poisonous alkaloids that are used in medicine. When infected rye is harvested the ergots may be grounds up into the flour made from the grain, and outbreaks of the disease known as ergotism may result from eating bread made from contaminated flour. Ergotism has two forms: it

FIGURE 17

The ergot fungus (Claviceps purpurea). *A, ear of infected rye plant, showing sclerotia (shaded black); B, sclerotia (ergots); C, sclerotium germinating to produce a stroma. All somewhat magnified and not drawn to the same scale.*

may affect the central nervous system, or it may cause gangrene. Both forms are extremely unpleasant, and the death rate is high. Modern systems of milling have greatly reduced the incidence of ergotism, but outbreaks still occur; there was quite a serious outbreak in 1951 in the village of Pont-St. Esprit in France which was traced to ergot-contaminated wheat. Ergot is also a cause of serious disease of cattle, who may feed on infected grain, or graze on infected fields.

One of the physiological effects of mild doses of ergot is

that it causes involuntary muscle to contract strongly. This is the basis of its medical uses, as a drug for administration during childbirth to increase the contractions of the womb, and still more as a drug for securing abortions.

Another interesting member of the Clavicipitales is *Cordyceps*, species of which attack insect larvae. The mycelium of the fungus fills the body of the insect without destroying its appearance, so that the insect becomes mummified. Club-shaped stromata of the fungus, which may be of considerable size, then grow out of the mummified body. *Cordyceps militaris*, a not uncommon species, parasitises caterpillars; its stromata stand up as small crimson clubs among grass in damp lawns or in hedgerows. *Cordyceps sinensis* was extensively used by the Chinese as a drug, possibly because of the current belief that it was a herb in summer and a worm in the winter. *Cordyceps norvegica* has stromata up to 20 centimetres (8 inches) long. *C. ophioglossoides* and *C. capitata* are parasitic on species of *Elaphomyces*, fungi that grow underground.

9 Powdery mildews and green moulds

In the cup fungi (Discomycetes) the ascocarp is usually a cup-shaped apothecium, while in the flask fungi (Pyrenomycetes) it is a flask-shaped perithecium. We now come to a third class of Ascomycotina where the ascocarp is more or less spherical and for the most part closed, without any opening for the emission of the spores. These fungi constitute the class Plectomycetes, and their enclosed fruit body is called a cleistothecium.

The Plectomycetes differ from the Discomycetes and the Pyrenomycetes in the shape of their asci, which are usually globose or broadly club-shaped, instead of cylindrical as in the other two groups. The asci of the Plectomycetes tend to be scattered in a haphazard manner within the cleistothecium, whereas in the Discomycetes and Pyrenomycetes they all arise at the same level on the ascogenous hyphae and so are arranged in parallel rows within the ascocarp.

The Plectomycetes contain several orders of fungi of which I am going to describe two: the Erysiphales or powdery mildews and the Eurotiales or green moulds.

THE POWDERY MILDEWS

The powdery mildews (Erysiphales) are an important order of plant parasites which do an immense amount of damage to crops and cultivated plants generally throughout the world. Most of them attack their host plants superficially, the mycelium spreading over the epidermis and anchoring itself by means of haustoria ('suckers') which it pushes into the epidermal cells of its host, thereby drawing nourishment. The name 'powdery mildew' comes from the vast number of conidia that the fungus produces on the surface of its host, so that the plant attacked is covered with a white powder of spores.

Two species of Erysiphales depart from the usual practice within the order by attacking their hosts internally instead of externally; these are *Leveillula taurica,* which parasitises a number of plants in the Mediterranean region, and *Phyllactinia corylea*, a parasite with a world-wide distribution. In *P. corylea* most of the mycelium is actually outside the host, but the fungus, which has no haustoria, sends hyphal branches into the host through stomata (ventilation pores) in the leaves; these hyphae come into close contact with the cells inside the leaf and draw nourishment from them.

The powdery mildews are obligate parasites, and have not yet been grown successfully on artificial culture media, although some of them have been grown on discs cut from leaves and placed in water or in a culture solution. *Erysiphe cichoracearum*, which is normally parasitic on plants of the cucumber family, has been cultured on tumour tissue.

The powdery mildews vary in their host specificity. At one time it was thought that they were nearly all confined to one genus of host plants, if not to one species, but it is now known that at least some of them are less host specific than this. *Erysiphe polygoni*, for instance, has been found attacking no less than 352 different species of host. At the other end of the scale, *Sphaeotheca phytoptophila* attacks only the galls produced by the mite *Phytoptus* on the western hackberry (*Celtis occidentalis*).

Some extremely serious plant diseases are caused by the powdery mildews. *Uncinula necator* causes powdery mildew of the grape vine, a disease that in the middle of the eighteenth century threatened the extermination of the wine grape in France until it was discovered that the fungus could be kept in check with sulphur. *Sphaerotheca pannosa* is the cause of powdery mildew of roses, which every gardener knows to his cost, and S. *mors-uvae* produces a powdery mildew of the gooseberry.

The general features of the life history of the powdery mildews are much the same for most species. Soon after the infection of the host the mycelium of the fungus forms a loose network spreading over the surface of the leaves, and the haustoria penetrate the epidermal cells and absorb nourishment for the fungus. As soon as the mycelium is established numerous conidiophores are produced, and these bear conidia which are usually formed in chains, though in some species the conidia drop off the conidiophores one by one as soon as they are formed. It is this conidial stage that produces the powdery appearance from which the powdery mildews get their name (Fig. 18a).

Later in the season the ascocarps are formed. The fertilisation

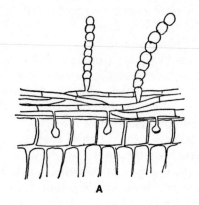

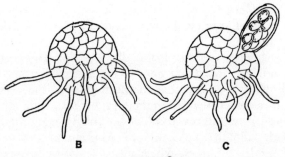

FIGURE 18

Powdery mildew of the rose (Sphaerotheca pannosa). *A, mycelium on the surface of a rose leaf, with haustoria in the epidermal cells of the host and chains of conidia; B, cleistothecium; C, cleistothecium bursting, with the single ascus emerging. All greatly magnified.*

of an ascogonium by an antheridium has been observed in several of the powdery mildrews, and, though mycologists disagree about the details of the sexual process, it seems safe to assume that asci are formed in the usual way. The asci are enclosed in small, spherical cleistothecia from the surface of which a number of appendages stick out. These may be of various kinds, depending upon the genus of the fungus. In *Erysiphe* and *Sphaerotheca*, for instance, the appendages resemble hyphae projecting from the surface of the cleistothecium (Fig. 18b); in *Uncinula* the tips of the appendages are coiled, in *Podosphaera* the appendages have branched tips, while in *Phyllactinia* they are stiff, resembling needles, with bulbous bases.

THE GREEN MOULDS

The green moulds belong to the order Eurotiales, which includes the two extremely common genera *Aspergillus* and *Penicillium*. Species of *Penicillium* are extremely common, forming more or less circular colonies, green with a white border, on decaying fruit, damp bread, leather, or almost any other kind of organic matter. *Aspergillus* may be found in similar places; its colonies are often green or blue-green like those of *Penicillium,* but may be yellow, brown, reddish, or even black, according to the species of Aspergillus concerned.

Although most species of *Aspergillus* and *Penicillium* are saprobes, a few are weakly parasitic, and some can cause diseases of man.

Penicillium is well known as the fungus from which the life-saving drug penicillin is obtained. Its mycelium consists of colourless, septate hyphae; the green colour of the colonies is due to the colour, not of the fungus, but of the spores, which are formed in long chains on the ends of conidiophores which are branched somewhat after the manner of the fingers of a hand. The arrangement resembles one of the flat brushes used by artists, and is called a penicillus, the Latin word for an artist's 'pencil', or brush (Fig. 19).

The chains of conidia in *Penicillium* are produced by special cells called phialides at the ends of the branches of the conidiophore; the cells of the branches immediately below the phialides

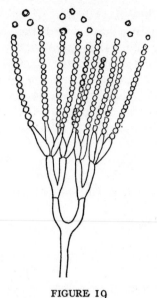

FIGURE 19

Tip of conidiophore and conidia of Penicillium, *showing metulae, phialides, and chains of spores. Greatly magnified.*

are called metulae. Phialides are of common occurrence in the higher fungi; they are cells, usually more or less flask-shaped, the narrow part or 'neck' of which has no cell wall at its tip; the protoplasm oozes out, rather like tooth paste squeezed from a tube, and rounds off, forming a conidium. Often several conidia are formed in succession from the same phialide, so that a chain of spores may be formed, as in *Penicillium*.

The conidia of *Penicillium* are usually coloured green, giving the colonies of the fungus a green colour, but in a few species they may be of some other colour. The white border of the colony consists of young mycelium which has not yet begun to sporulate.

Very few of the many species of *Penicillium* reproduce sexually; *Penicillium* is usually regarded as one of the Deuteromycotina for this reason. The few species that do undergo sexual reproduction produce rounded cleistothecia typical of the Plectomycetes.

ANTIBIOTICS

Penicillium is an extremely common mould and is responsible for a great deal of mould spoilage of food, textiles, and other things. In spite of these baleful activities, however, it must be regarded as the most valuable of all the fungi on account of the antibiotic penicillin that is obtained from it.

The discovery of penicillin came from one of those lucky accidents that occasionally illuminate the course of science. Just as the discovery of radium could be said to have started by the accidental fogging of some photographic plates stored in a laboratory cupboard, so the story of pencillin begins with the accidental contamination of a plate culture of bacteria. One day in 1929 Sir Alexander Fleming – then plain Dr Fleming – was working on some petri dish cultures of *Staphylococcus aureus*, a bacterium that gives rise to a number of septic infections of man. One of the plates showed a small round colony of a fungus, later identified as *Penicillium notatum*. The contaminated culture was useless as far as Fleming's work was concerned, but before throwing it away Fleming happened to notice that, round the colony of *Penicillium*, there were no bacteria growing. It seemed that something was diffusing out of the mould colony and killing the bacteria. This was new and, to a man of Fleming's calibre, very exciting.

Fleming decided to follow up his new discovery. He isolated the mould into pure culture, and tried the effect of using it deliberately to contaminate pure cultures of *Staphylococcus aureus*. His hopes were realised: in every case the mould killed the bacteria. Fleming then tried the effect of the mould on other bacteria. He found that wherever the bacteria were of the type called 'Gram-positive', reacting to certain blue stains such as methyl violet in such a way that the stain was not washed out by alcohol following treatment with iodine, the mould killed the bacteria. On 'Gram-negative' bacteria (blue stain washed out by alcohol after treatment with iodine) the mould had no effect.

It was now clear that the mould produced some substance that was highly fatal to Gram-positive bacteria. Fleming called this substance penicillin.

Fleming did not attempt to apply penicillin to the curing of disease, though he did use it in the laboratory for the isolation of Gram-negative bacteria from mixed cultures. By treatment of the cultures with penicillin he could kill the Gram-positive bacteria, leaving himself with only the Gram-negatives to sort out. In those days the amount of penicillin available for use was minute, and its medical use had to await the establishment of techniques for preparing it in larger quantities.

The study of penicillin was then taken up by Dr Harold Raistrick at the London School of Tropical Medicine. He made important contributions to our knowledge of the conditions that favoured the production of penicillin by *Penicillium notatum*, and he also found out something about its chemical nature. The yield of penicillin remained small, however, and this, combined with the indifference of the medical profession as a whole to the possibilities that the new drug offered in the fight against disease, led Raistrick to abandon the work. One is tempted to contrast this early attitude of doctors with what happened twenty years later, when penicillin became readily available; for a time it was prescribed for almost every ailment from septicaemia to gout, regardless of whether it was really indicated or whether the wretched patient happened to be allergic to it, as some people are. Fortunately, this enthusiasm has now waned to reasonable proportions.

After this, penicillin remained an interesting scientific curiosity until 1939, when a team was set up at Oxford under the direction of Dr H. W. Florey and Dr N. G. Heatley to work on its possible medical use. The outbreak of war was a stimulus to the work, for it became clear that, if penicillin was indeed any good, it would soon be needed for the treatment of battle casualties. For the first time penicillin was produced in sufficient quantities for the treatment of a few actual cases of human infection, and the results that were obtained were described as miraculous. It at once became clear that penicillin was a therapeutic agent of phenomenal power.

The exigencies of war, and the start of the 'blitz', began to interfere with research, so it was decided to move the work on penicillin to America, where there were facilities and manpower to spare. With the help of the Rockefeller Foundation Florey and Heatley were uprooted from Oxford and set down at the Northern Regional Research Laboratories of the Bureau of Agricultural Chemistry and Engineering at Peoria, Illinois. In 1941 there began what was to become one of the great classics of Anglo-American co-operation, and both nations can justly feel proud of what was accomplished at Peoria, Illinois, during the darkest days of World War II.

The most urgent need was to devise a means of producing penicillin in workable quantities. With the crude methods available at the time it needed the penicillin from 120 litres of culture fluid to treat even a minor infection, and serious cases needed far more than this. The penicillin was obtained by growing *Penicillium notatum* in a liquid culture medium and extracting the penicillin from the spent medium after straining off the mycelium. The culture vessels were quart milk bottles, and as the fungus would only grow on the surface of the medium the bottles could only be filled to a height of a few centimetres. This meant that, for every litre of culture fluid containing penicillin for extraction seven or eight milk bottles had to be used. The treatment of one case needed the penicillin from several thousand bottles.

The work at Peoria was started with *Penicillium notatum* cultures descended from the original fungus isolated by Fleming in 1929, but if this fungus produced penicillin, there was no reason not to suppose that other species of *Penicillium,* or other strains of the same species, might be better penicillin producers than the original isolate. Moreover, if one could be found that could be grown in deep culture – that is, one that would produce its mycelium throughout the depth of the culture medium instead of only on the surface – then it would be possible to produce penicillin in great fermentation tanks holding thousands of gallons of fluid instead of in milk bottles. This would revolutionise penicillin production overnight.

The penicillin workers sent out an urgent S.O.S. for any piece of rubbish bearing a colony of a green mould to be sent to the laboratory for examination. The American populace responded with a will, and for a time it looked as if the laboratories at Peoria might be sunk without trace beneath the ocean of mouldy bric-a-brac, from orange peel to old boots, that descended on them. Not content with this, the researchers appealed to Air Force stations overseas to send in soil samples for examination. *Penicillium* is a common soil fungus, and it might have been that the strain they so desperately needed could be found in the soil in Alaska, Sarawak, or the Phillipines.

Such industry did not go unrewarded for long, and the sample that hit the jackpot was not an exotic culture from the ends of the earth but an unromantic piece of mouldy squash peel from Peoria market. This yielded an isolate of *Penicillium chrysogenum* that produced penicillin in deep culture. True, the yield of penicillin was only about half the quantity per litre that *P. notatum* could provide, but this did not matter. The milk bottles were cheerfully thrown away, and in their place arose great fermenter tanks, each with a capacity of forty thousand litres or more. Soon there was penicillin to spare for every need.

This was only the beginning. The new isolate of *Penicillium chrysogenum* yielded 100 units of penicillin per millilitre of culture fluid (a unit of penicillin is defined as the equivalent of 0.0000006 grams of sodium penicillin G). This was extremely good when compared with the four units per millilitre obtained in the old Oxford days, but it was not good enough. The workers at Peoria set out to improve on this in order to satisfy their greed for penicillin.

First they tried to improve the yield by selection. A 'pure' culture of *Penicillium* is really a mixture of mycelia derived from thousands of spores. By isolating single spores and growing cultures from them individually it was found possible to select individual strains whose yield of penicillin was better than the average, and in this way the yield was raised to 250 units per millilitre.

The next thing to try was mutation. A mutation is a genetic change in an individual, owing to a change in one of the genes that control heredity. Mutations occur naturally, but when a certain mutation is being sought it is possible to speed up the naturally slow process by various artificial treatments, such as ultra-violet irradiation or exposure to X-rays. This was attempted at several research institutions in America, and as a result of work at the University of Wisconsin a new strain of *Penicillium chrysogenum* was obtained which yielded 900 units of penicillin per millilitre.

Nowadays the manufacture of penicillin and other antibiotics forms the basis of a vast industry, in which hundreds of millions of pounds are tied up. A modern antibiotics factory has a fermenter hall housing a dozen or more fermenter tanks, each with a capacity of up to eighty thousand litres. The fermenters are backed with big research departments in which large numbers of research workers are constantly testing fungi for antibiotic activity, besides working out improvements to existing techniques for the manufacture of antibiotics by fermentation, as well as their subsequent extraction from spent culture media.

Penicillin is in reality a group name for several closely allied substances, all derivatives of penicillanic acid. The different forms of penicillin vary in their antibiotic activity; the form most commonly used in medicine is known as penicillin G. Besides the natural penicillins we now have available synthetic forms made by the chemical treatment of penicillanic acid, which is itself prepared by fermentation. The value of these variations on the same theme is twofold. It has been found that some bacteria have evolved a resistance to the action of penicillin G and certain of the other forms of penicillin so that they are no longer effective; there is, however, at least a chance that these bacteria may be susceptible to the action of one or other of the synthetic penicillins. Also, some patients are allergic to some form of penicillin and so cannot avail themselves of it; in many cases a synthetic penicillin can be given without harm when a dose of penicillin G might have serious consequences.

The number of known antibiotic substances now runs into hundreds, if not into thousands. Few are of any medical use for one reason or another; they may be too highly toxic to be safely administered, for it is useless to kill the bacteria causing an infection if the patient also dies. Antibiotics which pass muster as regards toxicity may fail to kill bacteria as effectively as antibiotics already in common use, or their preparation or purification may be too troublesome or too costly to make them commercially valuable.

The Actinomycetes (organisms intermediate between fungi and bacteria) have proved particularly useful in the preparation of antibiotics by fermentation. Streptomycin, the next important antibiotic to be discovered after penicillin, has been a great help in the treatment of tuberculosis. Streptomycin will attack Gram-negative bacteria, of which *Mycobacterium tuberculosis* is one, whereas penicillin is only effective against Gram-positive organisms. Streptomycin is a little more toxic to man than penicillin, so its administration needs careful medical supervision. It is obtained from the actinomycete *Streptomyces griseus*.

Nowadays attention has turned towards the so-called broad spectrum antibiotics: antibiotics that are effective against a wide range of different kinds of organisms. Aureomycin is an example. Aureomycin is effective against Gram-positive and Gram-negative bacteria, Rickettsiae (very small bacteria), the larger viruses, and some Protozoa (minute parasitic animals). Aureomycin is a product of *Streptomyces aureofaciens*.

ASPERGILLUS

Aspergillus is a fungus closely resembling *Penicillium* in its general characteristics, but differing from it in the manner in which it produces its conidia. The tip of the conidiophore in *Aspergillus* is swollen to form a spherical vesicle, on the surface of which the phialides bearing the conidia are carried; the chains of spores stand out round the head of the conidiophore, giving it something of the appearance of the head of a mop (Fig. 20), from which the fungus gets its name (Latin *asper-*

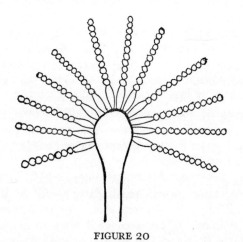

FIGURE 20

A conidial head of Aspergillus, *showing phialides and spores. Greatly magnified.*

gillum, a small mop used for distributing holy water). In some species, including the very common *Aspergillus niger,* the tip of the conidiophore bears short branches (metulae) on the ends of which the phialides are carried.

Many species of Aspergillus reproduce sexually, producing small, globular cleistothecia containing asci. Strictly speaking, those species of *Aspergillus* which do not produce cleistothecia should be classified in the Deuteromycotina, while those that do are placed in the Ascomycotina under the name *Eurotium.* There is a growing tendency, however, to retain the name *Aspergillus* for all species, whether they are known to reproduce sexually or not. Some authorities, in fact, call the order to which the fungus belongs the Aspergillales instead of the Eurotiales which, after all, is logical if you are going to drop the name *Eurotium* completely.

Some species of *Aspergillus,* especially *A. niger,* are used commercially for the production of various substances by fermentation, the most important of which is citric acid, used a great deal in making pharmaceutical preparations, in the food industry for flavouring and for making soft drinks, and for many other things including dyeing, calico printing, silvering, engraving and the manufacture of ink. Before 1923 about ninety per cent of the world's supply of citric acid came from Italy, where it was extracted from the juice of citrus fruits. Then a factory was set up in New York for the manufacture of citric acid by fermentation, and from this the fermentation process has developed into a considerable industry and the price of citric acid has fallen to a tenth of its former cost.

Aspergillus niger is used for the citric acid fermentation, sucrose (cane or beet sugar) being the usual substrate fermented. The fermentation is usually carried out in shallow aluminium trays, though deep fermentation methods are now coming into use.

Other organic acids are also manufactured by fermentation, using species of *Aspergillus.* Gluconic acid can be obtained by the use of *A. niger,* while for producing itaconic acid *A. terreus* is preferred. Various species of *Aspergillus* are also used in the preparation of gallic acid by fermentation. *A. ochraceus* is used in preparing cortisone from progesterol, and other species are used in a number of steroid transformations. The steroids are compounds chemically related to the steroids found in fats and oils. Some of the steroids are hormones of great physiological importance; testosterone, for instance, is one of the male sex hormones. Cortisone has an important medical use in the treatment of certain forms of rheumatoid arthritis, and other steroids are attracting more and more attention nowadays for their possible therapeutic value. Many steroids are extremely difficult to prepare by chemical means, so the fact that progesterone, easily obtained from plant products such as diosgenin, can be converted into a number of different steroids by mould fermentation is an exciting development in industrial microbiology.

10 The yeasts

It may surprise you a little to hear that yeast is a fungus, for it is not everybody's idea of what a fungus ought to look like. If you examine a speck of yeast teased up in a drop of water under a microscope, however, you will find that it consists of a mass of very small, colourless, more or less globular cells (Plate 11). The yeasts – there are many different kinds – belong to the Hemiascomycetes, the most primitive class of the Ascomycotina.

In addition to the true yeasts, there are many fungi that are yeast-like in that they have no mycelium of hyphae, but consist instead of single spherical or ovoid cells, sometimes known as a 'sprout mycelium'. There are also some fungi which have a yeast-like stage in their life histories into which they lapse under certain conditions. Such fungi differ from the true yeasts in that they do not form ascospores; they are called asporo-genous yeasts and are classified among the Deuteromycotina. They are often loosely called 'yeasts'. In this chapter I shall be mainly concerned with the true yeasts, of which the brewers' (or bakers') yeast, *Saccharomyces cerevisiae*, is a typical example.

The cells of the brewers' yeast are round or oval in outline and extremely small, measuring only about 10μ or so in diameter. Each cell is surrounded by a thin cell wall, and contains protoplasm and a central vacuole filled with fluid. To one side of the vacuole is the nucleus. For many years there was argument among mycologists about the nature of the yeast nucleus, many maintaining that the vacuole and the nucleus (then called the 'dense body' because nobody was quite sure what it was) together constituted a sort of compound 'nuclear apparatus'. In recent years photographs taken with the aid of the electron microscope, which has far greater resolving power than the optical microscope, have shown that the nucleus is surrounded by its own membrane and is quite separate from the vacuole, so there seems to be no reason for doubting that a yeast cell has a nucleus like that of any other fungal cell. There is still some doubt about the precise manner in which the yeast nucleus divides, but that does not concern us.

Besides the nucleus and vacuole, a yeast cell usually contains some small granules of a substance called glycogen, which is the carbohydrate most commonly found as a reserve food in the cells of fungi. The cell may also contain tiny globules of fat, especially if it has been grown on a medium poor in nitrogen and rich in carbohydrate (the same is true, unfortunately, of human beings who have been fed on a similar diet). Careful staining with basic dyes shows that the yeast cell contains plenty of mitochondria – minute bodies, almost at the limit of micro-scopical vision, which are concerned with certain biochemical activities in the cell.

Yeast cells show a peculiar method of vegetative reproduc-tion, called budding. A small protrusion, or 'bud', grows out of the side of a yeast cell and gradually increases in size; when it is about half the size of the parent cell it breaks away and starts an independent existence, the nucleus of the yeast cell having divided and one of the daughter nuclei having moved into the growing bud. (Fig. 21a). Sometimes the bud fails to separate from the parent cell, so that short chains of cells may be formed.

In some yeasts vegetative reproduction takes place differently, one cell elongating and then dividing transversely into two, after which the two cells separate from one another (Fig. 21b). Yeasts that do this are known as fission yeasts, most of them belonging to the genus *Schizosaccharomyces*.

When the brewers' yeast reproduces sexually a conjugation tube forms between two neighbouring yeast cells and the nucleus of one cell passes into the other and fuses with its nucleus. The cell in which the nuclear fusion has occurred then continues to reproduce asexually by budding. Some time later, after many generations of reproduction by budding, spore formation takes place, the contents of the yeast cells dividing up into four spores. These are set free by the dissolution of the wall of the mother cell, and each spore grows into a yeast

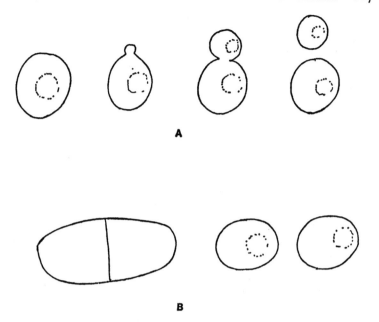

A

B

FIGURE 21

Vegetative reproduction in yeasts. A, by budding, as in most yeasts; B, by fission, as in the fission yeasts. Greatly magnified.

cell and reproduces by budding for several generations, until it is time for sexual fusion to occur again.

The four spores formed in a yeast are considered to be ascospores, the cell itself being the ascus.

This form of sexual reproduction is interesting when we come to consider its effect on the chromosome numbers (see page 73). The ascospores, like all other ascospores, are haploid, and when they begin to reproduce by budding they give rise to a race of haploid yeast cells. Then nuclear fusion occurs, and from the diploid cell so produced arises a strain of diploid yeast cells that again reproduce by budding. After an indefinite number of diploid generations produced by budding meiosis occurs and ascospores are formed, giving rise to the haploid strain again. When you look at cells of the brewers' yeast, therefore, you may be looking at haploid or diploid cells. A yeast which has both haploid and diploid strains in its life history is called a haplodiplobiontic yeast.

Yeasts do not all behave in this way. In *Schizosaccharomyces octosporus* conjugation takes place between two cells and is immediately followed by meiosis and ascospore formation, so that all the cells are haploid except the one in which nuclear fusion occurs. *S. octosporus* is a haplobiontic yeast. On the other hand, in *Saccharomycodes ludwigii* as soon as the ascospores are formed they undergo conjugation, forming diploid cells, so that all the cells in the life history are diploid except the ascospores. *S. ludwigii* is a diplobiontic yeast.

The greatest importance of the yeasts lies in their biochemical activity, and particularly in their power of fermenting certain hexose sugars such as glucose to form alcohol and carbon dioxide. It is this that underlies their use in brewing, baking, and wine making.

Alcoholic fermentation by yeasts is a form of respiration. Most plants, including fungi, respire by oxidizing hexose sugar to carbon dioxide and water, a process that can be represented (very inadequately) by the chemical equation

$$6C_6H_{12}O_6 + 6O_2 = 6CO_2 + 6H_2O$$

Since this requires the presence of oxygen, it is called aerobic repiration. The oxidation of the sugar releases energy, which can be stored up in 'energy rich' phosphates and used as required to supply the needs of the organism. The amount of energy obtained from a given amount of sugar is the same as would have been dissipated as heat if the same amount of sugar had been burned in air.

The equation quoted above represents the whole process of aerobic respiration from start to finish, but it is far more complicated than the equation indicates. Sugar is not directly oxidised to carbon dioxide and water; instead, it is broken down step by step by a number of enzymes working one after another. The first section of the process is known as glycolysis. This begins with the formation of sugar phosphate, which is then broken down by a number of different stages to an organic acid called pyruvic acid. In aerobic respiration the pyruvic acid then enters a complex cycle of chemical processes, called the Krebs cycle, as a result of which carbon dioxide is liberated and the oxidation of the original sugar is completed.

There is, however, another course the respiration can take (actually there are several, but only one of them concerns us here). Instead of entering the Krebs cycle, the pyruvic acid can be further broken down to alcohol and carbon dioxide, so that the equation for the breakdown of sugar reads like this:

$$C_6H_{12}O_6 = 2C_2H_5OH + 2CO_2$$

You will notice that no oxygen is required for this process, which is therefore called anaerobic respiration.

The amount of energy set free in anaerobic respiration is only about one-twentieth of that made available by aerobic respiration from a given amount of sugar. Few organisms are able to make use of the anaerobic process, but among those that can are some of the bacteria and the yeasts.

BREWING

The brewing of beer sounds simple in outline, but it is really an exceedingly complex process demanding a great deal of skill. No two brewers brew precisely the same beer, yet the product of any individual brewer must not be allowed to vary, or the customers will complain.

Basically, four main ingredients are required by the brewer: malt, hops, yeast, and water (known to the brewer as 'liquor'). A grist is made up from malt and other things, this is boiled with water, at which stage the hops are added, the product, known as wort, is fermented with yeast, and the resulting beer is subjected to certain 'finishing' processes before being bottled, or sent out in barrels of other containers for consumption by the thirsty populace.

Malt consists of grains of barley (sometimes other cereals such as wheat) that have been encouraged to germinate and then had their germination arrested by heating in a kiln. Cleaned barley grain is first steeped in water for one to two days, after which it is drained and either spread out on the floor of the malting house or placed in revolving drums to germinate. The growth promoting substance gibberellic acid is often added to the steeping water to reduce the time needed for germination, and also to reduce the amount of barley lost during the malting process. The gibberellic acid probably acts by stimulating enzyme activity in the grain.

During the germination of the barley an enzyme (actually an enzyme complex containing several enzymes) called amylase in the grain is activated and begins to convert the starch in the grain into a sugar called maltose. Before the conversion can proceed very far, however, the barley grain is dried by heat; this checks the conversion of starch into maltose but does not destroy the amylase in the grain, which is now known as malt.

The brewer's grist consists mainly of malted barley, but will probably contain other grains or grain products as well, the nature and amount of the added substances depending on the particular brewery, and the kind of beer that is being brewed. The grist is placed in a large vessel called the mash tun with water heated to a temperature of about 65°C (150°F), and the process called 'mashing' begins. The temperature of the water must necessarily be something of a compromise. Besides starch, the malt contains protein which is broken down to simpler substances during the mashing process. The degradation of protein is carried out by enzymes, called proteases, in the malt, and the optimum temperature for their action is around 40°C (104°F). The breakdown of starch by amylases, on the other hand is favoured by a higher temperature. However, the complex starch molecule consists of amylopectin and amylose, which are attacked by different enzymes, known as α- and β-amylases respectively, and α-amylase works best at a slightly higher temperature than β-amylase. One temperature, therefore, cannot suit all processes.

The water used in the mashing has a considerable effect on the quality of the beer. Soft water containing no more than 250-500 milligrams per litre of calcium sulphate or its equivalent is suitable for brewing pale ales and bitters, while for mild ales and stouts a harder water is best. It is for this reason that Burton-on-Trent has become noted for its bitter, while in London the mild is better. Not that the brewer always takes what water nature gives him, for it is a simple matter to alter its mineral content by suitable treatment. Sometimes the mineral content of the water is altered to match that of the famous water from Burton-on-Trent, a process known as 'Burtonising'.

In the mash tun enzymes from the malt break down the proteins in the grist to peptones and amino-acids, and the starch to dextrins and maltose. The process is usually not allowed to go quite to completion, for a certain amount of dextrin in the beer is an advantage, especially in the brewing of pale ale and bitter, since it adds stability. In brewing mild ale this does not matter, as mild ale is intended to be drunk as quickly as it is brewed. The extent to which the protein in the malt is degraded is also important; the presence in the beer of the more complex nitrogen compounds improves the flavour and also stabilises the foam.

When mashing is completed the solution, known as wort, is filtered through the solid residues in the mash tun and passed on to another large vessel called the copper. The 'goods' left behind in the mash ton are washed with fine jets of water, an operation known as sparging, to extract all the soluble matter, the washings being added to the wort in the copper. The solid matter left in the mash tun is usually disposed of as cattle food.

The wort in the copper is now boiled for about two hours. This has several effects. It stops the action of the enzymes from the malt, so that further breakdown of carbohydrates and proteins is halted. It also partially sterilises the wort; this is important before fermentation, for if the wort were to go into the vat as a sort of microbiological Noah's Ark the only prediction one could make about the result would be that the product would taste horrible. Boiling in the copper does not ensure sterility, but at least it sees to it that the yeast has a head start over possible competitors in the form of bacteria and wild yeasts (a 'wild' yeast is a naturally occurring species of yeast other than *Saccharomyces cerevisiae* that might get into the wort by accident).

The hops are added to the wort in the copper, and the boiling helps to extract the resins on which their value depends, though it also helps to destroy some of them. Hops are the female inflorescences of the hop plant (*Humulus lupulus*). The bracteoles that subtend the female flowers are covered with glandular hairs which secrete a number of different resins, collectively called lupulins. The resins from the hops impart a bitter flavour to the beer and also help to preserve it. The amount of hops added will depend on the type of beer that is being brewed; pale ales and bitters are more highly hopped than mild ales.

After boiling the wort is cooled and passed into the vat, where it is 'pitched' (inoculated) with yeast. The yeast used for pitching usually comes from a previous fermentation, and it is here that the wort is most likely to be contaminated with foreign micro-organisms, especially with yeasts such as species of *Rhodotorula*. Techniques of pure culture inoculation are now being devised to avoid this difficulty.

Yeasts can be divided into top and bottom varieties, according to whether they rise to the top of the vat during fermentation or sink to the bottom. Top yeasts are almost universally used for brewing English type beers, but for brewing lager beer a bottom type of yeast, such as *Saccharomyces carlsbergensis*, is preferred.

Fermentation in an open vat is usually carried out for five or six days, after which the wort has become beer. After fermentation the beer is treated with 'finings', such as isinglass or some other colloidal material, to clear it, and it is then 'racked' into casks or bottled.

For brewing lager beer a closed vat is used, and the fermentation is carried out for a longer period. After fermentation lager beer is stored in tanks for a period that may be as long as nine months before being bottled. The word lager comes from the German word *lagern*, to store.

Brewing by continuous fermentation is a new process that is being developed, and shows considerable promise for the future. Instead of brewing beer by the batch, the process is made continuous, wort going in at one end and beer coming out at the other, so to speak. The continuous brewing process still presents many difficulties, but the possibilities for the future are considerable.

WINE-MAKING

Wine-making is similar in principle to brewing: grape juice, containing sugar, is fermented with yeast, and the product is wine. Actually, almost any fruit or vegetable juice can be fermented into wine of a sort, but when one speaks of wine without any qualification it is understood that one means grape wine.

Ripe (but not over-ripe) grapes are crushed to express the juice, the crushing being carried out in some kind of mechanical press. The traditional method of crushing grapes by treading them has almost died out, though it is still done in a few of the sherry vineyards near Jerez; I hasten to add that the treaders wear special clogs on their feet, so the process is rather more hygienic than it sounds. The grape juice ready for the vat is known as must.

The must may or may not contain grape skins, and even stalks, in addition to the juice, depending on the type of wine. White wine may be made from white grapes or black, for the skins are removed before fermenting. Red wine is made from black grapes, the red colour coming from the pigment in the grape skins, which are included in the must. In the Bordeaux district the stalks of the grapes are usually removed, while in Burgundy stalks as well as skins go into the must. The inclusion of the stalks is not just laziness, for they contain a tannin which, besides modifying the flavour of the wine, help to preserve it.

The wine yeast is *Saccharomyces ellipsoideus*. Normally the must in the vat is not inoculated, as there are sufficient yeast cells floating in the air of the vat room, to say nothing of the yeasts that form a large part of the 'bloom' on the surface of the grapes, to start the fermentation. In some vineyards, however, it is becoming customary nowadays to inoculate the must with a pure culture *S. ellipsoideus*, and in America it is the usual practice.

Since the must is not boiled before being put into the vat it contains a considerable variety of micro-organisms, including wild yeasts, and some of these may give trouble. It is the usual custom to treat the must with sulphur dioxide before fermentation, either in the form of sulphur dioxide gas or as sulphite or metabisulphite. This must be done with care, for sulphur dioxide has an extremely unpleasant and penetrating taste which could ruin any wine. Sulphuring is especially dangerous with the more delicately flavoured white wines, and in a cheap white wine the taste of sulphur dioxide is usually detectable. Red wines, with their fuller flavour, disguise the sulphurous taste more effectively.

The idea behind sulphuring is that the wild yeasts are, fortunately, more susceptible to the poisonous action of sulphur dioxide than the wine yeast, so that they are selectively killed. Actually, there are usually enough wild yeasts left alive to start the fermentation of the wine. This does not seriously matter, as the wild yeasts are sensitive to alcohol, so that when the alcohol content of the wine rises to about four per cent the activity of the wild yeasts ceases and the wine yeast takes over. The wine yeast can tolerate alcohol concentrations of up to twelve per cent, but higher concentrations than this kill it.

After fermentation in the vat, the wine is usually run into tanks, where a further slow fermentation takes place, continuing until all the sugar is fermented. The new wine is then racked off, leaving the sediment behind, and aged in tanks which are completely filled and sealed, so as to exclude air. The wine is drawn off at intervals, and after from two to five years of this treatment it is finally put into containers for distribution. Sparkling wines such as Champagne are given a second fermentation in the bottle, a little sugar and yeast being added. The carbon dioxide evolved during this fermentation gives the wine its effervescence. Cheaper wines are given this second fermentation in closed fermenters. In very cheap 'wines' the effervescence may be added by blowing in carbon dioxide under pressure instead of by fermentation, but this is unsatisfactory as such wines quickly lose their sparkle when opened.

No yeast can live in a medium containing more than twelve per cent of alcohol. Wines such as port and sherry, the alcoholic content of which is about twenty per cent, are fortified by the addition of brandy after fermentation.

BREAD

In brewing and wine-making yeast is used to produce alcohol, but in baking it is used to provide carbon dioxide; the bubbles of carbon dioxide liberated in the dough cause it to 'rise' and give the finished loaf its porous consistency.

The bakers' yeast is the same species as the brewers' yeast, *Saccharomyces cereviseae*, a strain being chosen which has the properties required by the baker. The yeast is added to a mixture of flour and water, often supplemented by dough conditioners such as common salt and salts of ammonia, the latter serving as additional nitrogenous food for the yeast. Flour contains a certain amount of natural amylase, and this ferments some of the starch in the flour to maltose, which is then acted on by the yeast, which ferments it first to hexose sugar and then to alcohol and carbon dioxide. During the fermentation process bubbles of carbon dioxide cause the dough to expand to several times its original volume.

On baking the bread the activity of the yeast is stopped, and under the action of the heat of the oven the bubbles of carbon dioxide in the bread expand, increasing the volume of the loaf still further. Most of the alcohol generated by the yeast fermentation is lost during the baking, though a newly baked loaf may contain as much as 0.5 per cent of alcohol, which accounts for the pleasant yeasty smell and flavour of a new loaf. The characteristic flavour of home made bread is mainly due to the action of contaminating micro-organisms in the dough during the somewhat longer period that a home made loaf is given to ferment.

THE MANUFACTURE OF BAKERS' YEAST

The manufacture of bakers' yeast is a considerable industry. Strains of *Saccharomyces cerevisiae* are selected for their high respiration rate, so that they will produce carbon dioxide quickly in the dough. The usual practice is to transfer a single cell of the selected yeast to a tube of culture medium, and to use the yeast culture grown from this to inoculate a flask containing a larger volume of culture medium. This process is repeated several times, the amount of medium increasing with each transfer as the amount of culture transferred becomes larger, until finally a sufficient volume of culture is obtained to act as an inoculum for a production fermenter tank containing from 40,000 to 200,000 litres (10,000 to 50,000 gallons) of culture medium.

The yeast cultures are kept well aerated at all stages, to encourage aerobic rather than anaerobic respiration. It must be emphasised that yeasts respire aerobically, like other organisms, if there is enough oxygen available; they only resort to anaerobic respiration under conditions of poor aeration. Under aerobic conditions yeasts grow faster, producing a greater number of cells in a given time, which is the object of the yeast manufacturer.

Great care is taken at all times to avoid contamination of the yeast with foreign micro-organisms. The general state of asepsis in a yeast factory approaches that of a surgical operating theatre.

When the growth of the yeast is completed the culture is rapidly cooled, and the yeast cells are separated from the culture medium by a centrifuge and filtered in a filter press. The resulting mass of yeast cells is mixed with a plasticiser and cut into blocks which are wrapped and placed in cold storage.

OTHER USES OF YEASTS

Besides brewing, baking, and wine-making, the yeasts have several other uses, both major and minor. Formerly a great deal of industrial alcohol was prepared by yeast fermentation, but nowadays most of it is prepared chemically from petroleum. A certain amount is, however, still made from carbohydrate fermented by yeast, and alcohol prepared in this way is, for certain purposes, regarded as superior to alcohol obtained from petroleum.

Yeast cells contain vitamins of the B complex; they also contain ergosterol, a substance which, when irradiated with ultra-violet rays, produces vitamin D. Yeasts therefore have an important pharmaceutical use as sources of these vitamins, especially in these days when the public, thanks to television commercials, are inclined to be a little too vitamin-conscious!

Yeast is a nourishing food, for when dried it contains about fifty per cent of high quality protein. Unfortunately its excessively bitter taste has prevented it from becoming popular as human food, though during World Wars I and II it was used to a small extent by both sides to supplement their rations. Some of the asporogenous yeasts, such as *Torulopsis utilis*, are less bitter than the brewers' yeast and more suitable on that account for use as human food. The development of space travel has made this aspect of yeast important, for as concentrated food yeast would be hard to beat.

Animals are less fussy than man about what they eat (though some cat owners may not agree). Excess yeast from the breweries, therefore, finds an important market as a constituent of fodder for farm livestock.

Yeast is an important source of enzymes which can be extracted from the cells and used for a variety of different purposes. One little-known use of yeast, or of enzymes extracted from it, is in the making of chocolate creams. In order to apply the chocolate coating the centres must be firm, but in the finished product they are preferred soft and creamy. If a little yeast, or the enzyme invertase obtained from yeast, is mixed with the centre before coating, some of the cane sugar in the centre is converted into 'invert sugar', a mixture of glucose and fructose which is more soluble than sucrose, so that the centres become softened after they have received their chocolate coating.

11 The grey and white moulds

The grey and white moulds include the very common fungi popularly called 'pin moulds' because their fluffy growth resembles, under the miscroscope, a collection of round-headed pins stuck in a pin-cushion. The heads of the pins are the sporangia that contain the spores. The pin moulds belong to the order Mucorales of the sub-division Zygomycotina which, with the sub-division Mastigomycotina, make up the group known loosely as the lower fungi, in contrast to the Ascomycotina and Basidiomycotina which comprise the higher fungi.

The most familiar name among the pin moulds is *Mucor*. Any reader who studied biology at school will remember seeing a greyish mould growing on a piece of damp bread under a bell jar, and being told that this was *Mucor*; in actual fact it was probably nothing of the sort, for *Rhizopus stolonifer*, another of the pin moulds, is far more common than *Mucor* on damp bread. The best way to obtain *Mucor* is to place some fresh horse dung under a cover for a few days, when a good growth of one or more species of *Mucor* is almost certain to appear.

The name *Mucor* has been known for upwards of two and a half centuries. It was first proposed by an Italian botanist, Pier Antonio Micheli, in a book called *Nova Plantarum Genera*, published in 1729 in Florence. Micheli described his fungus as occurring on horse manure, on fragments of wheat stalks, and on semi-decomposed branches of conifers. Micheli described four different species of *Mucor*, but since his time the number has grown and there are now from forty to eighty species, according to your views on taxonomy.

As a genus, *Mucor* is widely distributed in nature. It grows luxuriantly on damp organic matter of many kinds, where it forms white, brown, or grey fluffy colonies. It is common in the damp litter on woodland floors, and also in soil, in fresh horse dung, and may occur on mouldy food of almost any

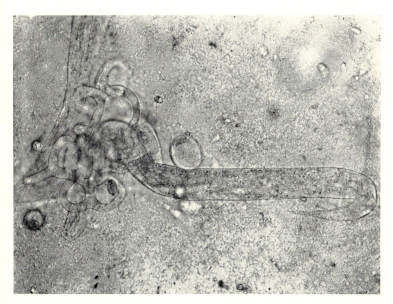

Plate 13. Photomicrograph of an eelworm captured by the sticky networks of *Arthrobotrys robusta*

Plate 14. Photomicrograph of two eelworms captured by the constricting rings of *Dactylaria gracilis*. The eelworm on the left has been recently captured and still retains its contents, while that on the right has been captured for about twenty-four hours and clearly shows the trophic hyphae inside its body. Greatly magnified

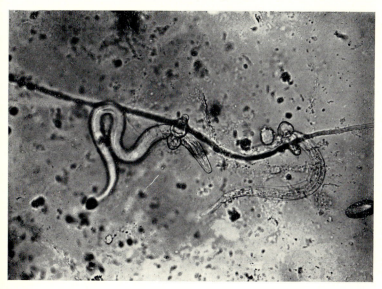

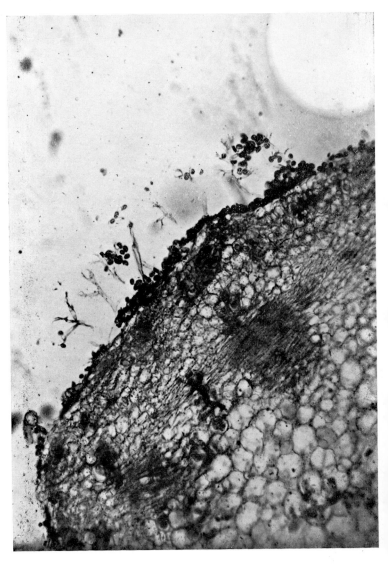

Plate 15. Photomicrograph of part of a transverse section of a stem attacked by *Peronospora*. The sporangiophores of the fungus can be seen jutting out from the surface of the stem. Magnified

kind, and mouldy fruit. When *Mucor* is growing on organic matter of any kind it produces a characteristic musty smell, from which it gets its name; its old Etruscan name was Mussa, derived from the Greek word Mephitis, signifying a foul odour.

The mycelium of *Mucor* consists of a weft of tubular hyphae which are normally without cross-septa dividing them into cells, though septa are formed here and there, mainly to isolate portions of old or dead mycelium; the reproductive organs are also regularly cut off from the hyphae by septa. This absence of septation is characteristic of the lower fungi, and is usually sufficient to distinguish them at a glance. The hyphae of *Mucor* contain protoplasm and numerous nuclei. A structure of this sort, where many nuclei control a common portion of cytoplasm, is called a coenocyte. The colour of the mycelium of *Mucor* may be white, grey, or brownish.

Mucor reproduces asexually by spores which, because they are produced in a structure called a sporangium, are known as sporangiospores. When asexual reproduction is about to begin numerous erect branches grow upwards from the mycelium; these are the sporangiophores that bear the sporangia. They are usually wider than the hyphae from which they arise. The tip of a sporangiophore swells, forming a globular sporangium which is cut off from the sporangiophore by a septum which is dome-shaped, arching into the sporangium; this is called the columella. The sporangium contains dense protoplasm which becomes cut up into a large number of spherical or, more commonly, elliptical spores (Fig. 22).

A young sporangium of *Mucor* is usually white, but as it ages its colour changes, passing through yellow and brown to black, dark brown, grey, or, in some species, olive green or even pink. Sometimes, as in *Mucor plumbeus* (*M. spinosus*), the wall of the sporangium is covered with crystals which usually consist of oxalates of calcium or other metals.

The sporangiophore may be unbranched and carry one sporangium only as in the well known *Mucor mucedo*, or it may be branched, with a sporangium at the end of each branch as in *M. racemosus*. Where the sporangiophore is branched the individual sporangia are somewhat smaller than the single sporangium of species with unbranched sporangiophores.

E

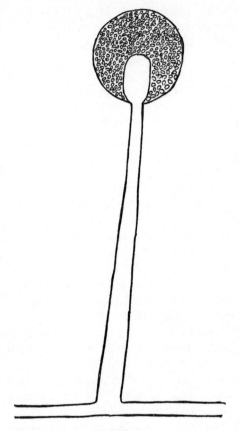

FIGURE 22

A sporangium of Mucor mucedo. *Note the absence of septa and the columella arching into the sporangium. Magnified.*

When the spores are ripe they are set free from the sporangium in one of two ways. The species of *Mucor* that is usually dealt with in school biology syllabuses is *M. mucedo*, and spore liberation in this species almost invariably described wrongly in school text-books. In this species, as indeed in most species of *Mucor*, the spores are slime spores: they stick together in a mass of slime. When the spores are ripe the wall of the sporangium dissolves in water coming from the sporangium and, possibly, from the surrounding air. This leaves the spores surrounded by a tiny globule of water, called the sporangial drop. This gradually dries, leaving the mass of spores firmly cemented round the columella. How they are finally dispersed is something of a mystery. They are too firmly stuck to be dislodged by wind; possibly they come unstuck when struck by rain drops, or they may be dislodged by insects.

The base of the sporangium adjoining the columella does not dissolve, but remains as a small frill round the tip of the sporangiophore, called the collarette. Presumably some chemical change takes place in the wall of the sporangium to render it soluble, but what this change is we do not know.

Most species of *Mucor* follow the pattern of spore liberation just described, but in a few species the procedure is different. These species have dry spores. The wall of the sporangium fragments at maturity, leaving the spores in a loose bunch round the columella, whence they are easily blown away by the wind.

Some species of *Mucor* produce a different kind of asexual spore in addition to the sporangiospores. A short section of a hypha becomes filled with particularly dense protoplasm and secretes a thick wall round itself, subsequently being released by breaking away form the rest of the mycelium. Such a structure is called a chlamydospore; owing to their thick walls, chlamydospores are resistant to adverse conditions, and their formation may help the fungus to tide over a bad period, such as drought or extreme cold, when life might be hazardous for the vegetative mycelium.

Sexual reproduction in *Mucor* takes place by the fusion of a pair of branches called gametangia (a gametangium is a structure containing sex cells or gametes, just as a sporangium is a structure containing spores). The process starts when two hyphae approach one another and lie side by side. From each hypha small swellings develop opposite to one another, from which branches grow out tip to tip, pushing the hyphae apart somewhat as they grow. In each branch a septum is formed, cutting off the gametangium from its stalk, which is called the suspensor. The walls between the two gametangia break down, so that their protoplasm mingles; this is plasmogamy. A thick wall is formed round the fused protoplasm, forming a rounded, dark-coloured structure called the zygospore (Fig. 23).

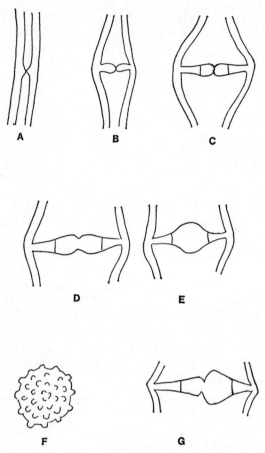

FIGURE 23

Sexual reproduction in the Mucorales. A – E, formation of the gametangia and sexual fusion in Mucor genevensis; *F, zygospore of* M. hiemalis; *G, fusion of gametangia in* Absidia, *the gametangia being of different breadths. Greatly magnified*

Both the gametangia are multinucleate at the time of fusion. After plasmogamy, while the young zygospore is increasing in size, nuclear divisions occur, still further increasing the number of nuclei present. Then nuclear fusions occur, and after about four days virtually all the nuclei have fused with an opposite number. The nuclei then shrink a little in size, and the zygospore passes into a state of dormancy which, in *Mucor genevensis*, commonly lasts about four months.

Between the fusion of the nuclei and the onset of the dormant period the nuclei in the zygospore appear to increase in number. This is probably the result of a reduction division (meiosis) which reduces the diploid chromosome number to the original haploid number, though this is not certain.

The details of sexual reproduction in *Mucor* was worked out by Cutter in 1942, using *Mucor genevensis* as his material. Other species of *Mucor* probably follow the same general pattern.

It is too often stated in school text-books that the gametangia of *Mucor* grow out and find each other by some kind of mutual attraction. This is nonsense. The *hyphae* from which the gametangia arise may show mutual attraction, growing towards each other until contact is made. The actual gametangia, however, are in contact from their inception.

Many species of *Mucor* are heterothallic, the two gametangia that unite during zygospore formation coming from opposite mating strains, plus and minus. In heterothallic species of *Mucor* such as *M. mucedo* and *M. hiemalis*, it is found that if a culture is raised from a single zygospore, the sporangia formed by it are all plus or all minus. This is not true for all the Mucorales; in *Phycomyces*, for instance, both plus and minus sporangia result from the germination of a single zygospore.

A few species of *Mucor* are homothallic. In these species, zygospores are formed even in cultures that have been raised from a single spore.

When the zygospores of *Mucor* germinate a short germ hypha emerges which develops directly into a sporangiophore bearing a sporangium. The spores from this germinate and form new mycelia.

In the sexual reproduction of *Mucor* both the pair of gametangia which conjugate are the same size and shape; there is no morphological distinction between male and female. These terms, in fact, can scarcely be applied to sexual reproduction of this kind, which is said to be isogamous. In some moulds of the order Mucorales the gametangia differ in form; in *Zygorrhinchus*, for instance, one is longer than the other, while in *Absidia* they differ, not in length, but in width (Plate 12, Fig. 23g). Sexual reproduction in these genera is said to be anisogamous. The spores formed by both isogamous and aniso-gamous sexual reproduction are called zygospores.

THE CAP-THROWER

The cap-thrower (*Pilobolus*) is a member of the Mucorales with an interesting method of spore dispersal. There are several species of *Piloblus*, of which *P. longipes* and *P. kleinii* are common inhabitants of horse and cow dung. The dispersal mechanism may easily be studied with a hand lens if a lump of fresh dung is placed in a moist atmosphere under a bell jar or a plastic bag; it is essential that it should be kept in the light, for the sporangia are not formed in the dark.

The sporangia of *Pilobolus* begin to appear in from three to ten days after the dung is gathered. The first intimation that sporangia are about to be formed is the appearance, usually in the morning, of tiny yellow bulbs, half buried in the dung; they can be seen with the naked eye, but careful inspection is needed since they do not differ greatly in colour from the surrounding dung. In an hour or two a yellow process will have grown up from each bulb, pointing towards the strongest light, another instance of the phototropism that I mentioned in Chapter 7 in connection with the asci of *Pyronema omphalodes*. At this stage the dung will present the appearance of being covered with a growth of hairs, each about a millimetre long, all pointing more or less in the same direction.

The sporangium itself usually forms during the night following the formation of the sporangiophore. The sporangium is elliptical in shape, with the top half of the ellipse coloured an intense black, and the top of the sporangiophore, just beneath the sporangium, is swollen to form an egg-shaped vesicle which is larger than the sporangium itself. The walls of the vesicle can be seen, under the low power of the microscope, to bear droplets of water. The sporangium is cut off from the tip of the sporangiophore by a dome shaped columella, as in *Mucor* (Fig. 24).

When the sporangium of *Pilobus* is about to be discharged a circular split occurs in the lower, colourless half of the

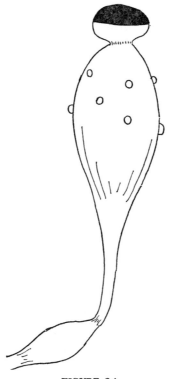

FIGURE 24

Sporangiophore and sporangium of Pilobolus. *Magnified.*

sporangium, near to its junction with the vesicle on the end of the sporangiophore. This rupture of the sporangial wall is caused by swelling of a mass of mucilage which is situated just below the spores. The mucilage bulges out through the split in the wall of the sporangium.

Before the split in the sporangial wall occurs the vesicle at the top of the sporangiophore is completely distended with water; measurements have shown that the pressure inside is about five atmospheres. As soon as the break occurs the wall of the sporangiophore, which is elastic, contracts. This squirts a drop of sap out of the end of the sporangiophore which flies to a distance of up to two metres, carrying the sporangium with it. If the dung on which the fungus is growing is situated in a field, the drop of water will most likely impinge on a blade of grass at the end of its trajectory. When the water droplet evaporates the sporangium will be left sticking to the grass, where it will be eaten by a herbivorous animal.

If this happens, the spores are released in the gut of the animal. Not only are they unharmed by the digestive processes going on around them : there is evidence that their powers of germination are increased by passage through the gut of the animal. In due course the spores are voided with the dung of the beast, whereupon they germinate to produce a mycelium in the dung with, in time, another crop of sporangia ready to be fired on to another blade of grass.

Apart from the sophisticated hydraulic adaptations to spore dispersal, the structure of the sporangium itself is adapted to a remarkable degree to ensure the effectiveness of the mechanism. The outer half of the sporangium is black and, like the back of a duck, unwettable; the lower half, on the other hand, with its projecting mass of mucilage, readily absorbs water. When the sporangium is flying through the air in its droplet of sap, there-fore, the black half of the sporangium projects from the surface of the droplet, while the mucilage is submerged. On impact the sporangium is for the moment submerged completely, but it immediately bobs up again with the black top projecting from the surface of the droplet and the mucilage submerged and, therefore, pointing towards the surface of the grass leaf on which the droplet has impinged. As the droplet evaporates the mucilage comes into contact with the leaf surface, to which it sticks.

In such a perfect arrangement one might suspect that even the blackening of the top of the sporangium would play a part in the general set of adaptations. This is indeed the case. It is thought that the blackening of the top of the sporangium serves to protect the spores from ultra-violet rays contained in sunlight, which might otherwise damage them.

SPORANGIAL EVOLUTION IN THE MUCORALES

The order Mucorales shows an interesting evolutionary line (more correctly, several lines) running through it, resulting in an increase in the number of sporangia born on one sporangiophore, with a reduction in the size of the sporangia as their numbers increase.

The starting point in the series is *Mortierella*, species of which are particularly common in the soil. Here each sporangiophore bears a single terminal sporangium of moderate size which is of an extremely simple type. The septum that divides sporangium from sporangiophore is not arched out into the sporangium to form a columella as in *Mucor*; it is a flat wall across the end of the sporangiophore.

The next stage of evolution is seen in *Mucor*. In *M. mucedo* the sporangiophore bears a single terminal sporangium which is extremely large, and the septum separating it from the sporangiophore arches upwards forming a dome-shaped columella. It is thought that the presence of a columella makes it easier for food material to pass from the sporangiophore into the sporangium to nourish the large number of spores.

In some species of *Mucor,* such as *M. racemosus,* the sporangiophore is branched, with a sporangium at the end of each branch, the individual sporangia being smaller than the single sporangium of *M. mucedo.* The sporangia in *M. racemosus* are provided with columellae. *M. racemosus* shows the first stage in the evolution of greater numbers of smaller sporangia.

The next stage in the evolution of the sporangia is shown by *Thamnidium.* Here there is a moderately large terminal sporangium provided with a columella, but in addition there are short branches from the sporangiophore just below the terminal sporangium which bear at their tips small sporangia, called sporangioles. The sporangioles are without columellae, and each contains only a few spores, which are often somewhat larger than the spores borne in the terminal sporangium. The sporangioles become detached bodily from the branches and are dispersed by air currents. When they finally come to rest water is absorbed, the spores swell, and the wall of the sporangiole is ruptured, setting free the spores.

In some species of *Thamnidium,* grown under special conditions, only the sporangioles are formed, the terminal sporangium being absent altogether.

In *Dicranophora* the sporangioles are smaller and more numerous, and they only contain one or two spores each. The terminal sporangia are often absent.

The line of evolution reaches its logical conclusion in *Chaeto-cladium*, where the many sporangioles are one-spored and indehiscent : that is, they do not open to release their spores, the spore germinating while still enclosed in the sporangiole. As in *Dicranophora*, the sporangioles are abstricted from the branches that bear them and dispersed by wind. Moreover, in *Chaeto-cladium* sporangioles only are formed; there is never a large terminal sporangium as well.

In *Chaetocladium*, with its wind-dispersed, indehiscent, one-spored sporangioles, the sporangiole has virtually become a conidium such as we find in the Ascomycotina. The only essential difference is that, if we examine the sporangiole with its single spore under the highest power of the microscope, we can distinguish two walls, the wall of the sporangiole and the wall of the spore inside it. In a true conidium there is, of course, only one wall, the wall of the conidium itself.

The genera described above show only one line of evolution in the sporangia of members of the Mucorales. Other lines exist. For example, in *Blakeslea* well nourished colonies produce large terminal sporangia with large columellae, while if the colony is half-starved the terminal sporangia are smaller, and do not have columellae. Instead of the terminal sporangium, sporangioles may be formed on branches from the tip of the sporangiophore. The sporangioles are formed in a different way from those of *Thamnidium*, *Dicranophora*, or *Chaetocladium*. The tip of the fertile branch is swollen to form a spherical vesicle the surface of which bears numerous sterigmata (short points that jut out) to which the sporangioles are attached. Instead of having one sporangiole per branch, therefore, each branch bears many sporangioles.

Continuing the line of evolution shown by *Blakeslea*, in *Choanephora* the sporangioles are one-spored and indehiscent as in *Chaetocladium*. Like *Blakeslea*, *Choanephora* may form a large terminal sporangium with a columella instead of sporangioles. Finally, in *Cunninghamella* the sporangioles are one-spored and indehiscent, and the large terminal sporangium is never formed.

A third line of sporangial evolution is found in the family Piptocephalidaceae. Here the sporangium is long and tubular, with the spores arranged in a single line. When the spores are mature the sporangial wall disintegrates, leaving the spores stuck end to end in chains, not unlike the chains of conidia in *Aspergillus*. In *Dispira*, *Piptocephalis*, and *Syncephalastrum* the tip of the sporangiophore is branched, while in *Syncephalis* it is unbranched but has a terminal swelling that bears the sporangia.

Lastly, in the family Kickxellaceae the sporangia are cylindrical, indehiscent, one-spored structures borne on sterigmata which are developed on special lateral branches of the sporangiophore, called sporocladia.

THE MUCORALES IN INDUSTRY

While the Mucorales do not achieve the industrial importance of *Aspergillus niger*, some of them are used for industrial fermentations, the genus *Rhizopus* being particularly useful. *Rhizopus stolonifer*, *R. arrhiza*, and other species of *Rhizopus* are used in the preparation of some of the steroids from progesterone. Species of *Rhizopus* are used in the production of enzymes, *R. stolonifer* provides us with fumaric acid, and *R. oryzae* is responsible for the production of lactic acid by fermentation.

12 Fungi that hunt

Some of the strangest and most exciting of all the fungi are the predacious fungi : fungi which capture microscopic animals alive and feed off their bodies, both while they are alive and when they are dead. The capture of a living animal, even one as slow moving as the amoeba, poses a problem for a fungus, and many of the predacious fungi aim higher than that, capturing eelworms as their prey, and an eelworm, though microscopic, is one of the most active animals alive.

The predacious fungi have evolved ingenious methods of dealing with this problem, the eelworm-trappers especially being provided with eelworm traps of considerable sophistication. Some trap eelworms by means of sticky networks in which the eelworms become entangled; others rely on mechanical traps, lassooing their prey with constricting rings that close about the eelworm's body and crush it with remorseless tenacity.

There are two main groups of predacious fungi. One, the Zoopagales, is an order of the Zygomycotina, allied to the Mucorales, while the other group belongs to the Deuteromycotina. The predacious habit is not confined to these two groups, however, for predacious genera crop up here and there in nearly all the major groups of fungi.

THE ZOOPAGALES

The Zoopagales have only been known for a relatively short time. In 1933 Drechsler, in America, described some curious fungi that captured amoeba and other small Protozoa in leaf mould, and two years later he erected the family Zoopagaceae to include this particular section of what are now called the predacious fungi. Since then, the family has grown to include about a hundred species and has been raised to the rank of an order, the Zoopagales.

There are two types of vegetative structure to be seen in the Zoopagales. In one type the fungus forms a mycelium of hyphae like that of a mould, while in the other there is no mycelium, the body of the fungus consisting of a coiled structure inside the body of the host. It is hard to believe that fungi with such very different kinds of body organisation belong to the same order, but their spores clearly indicate that they are related.

In the mycelial Zoopagales the mycelium consists of extremely fine hyphae, no more than 1 to 2μ in diameter in those species that capture Protozoa, though the hyphae of the few species that capture eelworms are coarser. The hyphae are branched, but the branching is sparse and they do not form the dense, fluffy growth that is characteristic of the Mucorales described in the last chapter. It is easy not to notice the mycelium of the Zoopagales when one is examining a mixed culture containing moulds of various kinds.

Most of the Zoopagales prey on the small amoebae that abound in soil and rotting vegetable matter. An amoeba is a microscopic animal of no definite shape. It moves by pushing out blunt processes, called pseudopodia, from its body, by means of which it can drag itself slowly over the terrain. Its most conspicuous feature is its contractile vacuole, a fluid-filled space, circular in outline, which is constantly expanding and contracting as the amoeba pumps excess water out of its body.

If an amoeba, in the course of its wanderings, happens to touch a hypha of one of the Zoopagales it sticks, held, presumably, by a sticky substance that the fungus secretes. When the prey has been secured the fungus puts out a fine haustorium into the body of the amoeba and begins to absorb its contents (Fig. 25). At first the amoeba seems but little perturbed by the presence of the haustorium in its body; the contractile vacuole continues to expand and contract, and the normal streaming movements of its protoplasm can still be observed. After a time, however, as its body contents are gradually depleted, it withdraws its pseudopodia and, assuming a more or less rounded shape, becomes motionless. At what particular instant after that the amoeba can be said to be dead is difficult to determine, but its body becomes increasingly transparent as its material is absorbed by the fungus, and finally nothing is left but the

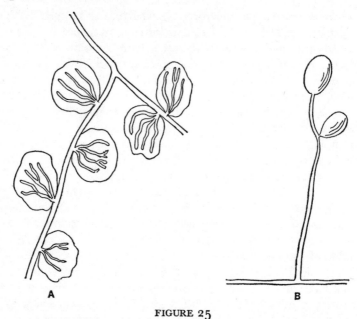

FIGURE 25

Stylopage cymosa. *A, amoebae captured by the mycelium, with haustoria absorbing their contents: B, a conidiophore bearing two conidia. Greatly magnified.*

shrivelled outer 'skin' or pellicle of the amoeba attached to the hypha of the fungus.

Other prey than amoebae may be captured by different species of Zoopagales. Several species capture testaceous (shelled) species of Rhizopoda, the group to which amoeba belongs, and a few, such as *Stylopage hadra* and *S. grandis,* can even prey on eelworms.

Eelworms, or nematodes to give them their scientific name, are microscopic roundworms; they are related to the 'worms' that sometimes infest the stomachs of cats and dogs, though far smaller, the eelworms found in soil usually measuring from 0.1 to 1 millimetre (1/250 to 1/25 inch) in length. They are extremely active creatures, moving with rapid undulations of their bodies, giving them an eel-like appearance under the microscope; hence their name. Most soil species are free-living, but some are important parasites of plants; every farmer and gardener has heard of the potato root eelworm, the depredations of which cost this country about £2,000,000 a year in lost potatoes.

To capture such an active prey as an eelworm a fairly robust mycelium is needed, so it is scarcely surprising that those Zoopagales that catch eelworms have hyphae of larger diameter than those that prey on amoebae *Stylopage grandis,* for instance, has hyphae measuring 5μ or more in diameter. Eelworms are captured by adhesion in the same way as amoebae, and with these larger hyphae the sticky fluid by which the eelworms are held can be clearly seen (Fig. 26). Unlike an amoeba,

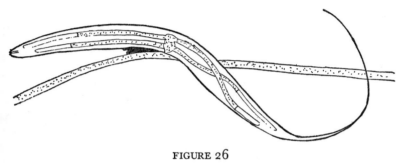

FIGURE 26

An eelworm captured by Stylopage grandis. *The sticky fluid holding the eelworm is shaded black. Greatly magnified.*

an eelworm is everything but a passive captive. When caught it struggles violently for freedom, dragging the hypha of the fungus back and forth with such force that a breakage seems certain. The fungus is stronger than the eelworm, however and it hangs on grimly until, after an hour or so, the captive ceases to struggle.

What actually causes the death of the eelworm is by no means certain. The idea that it dies of fright can be dismissed as frivolous; similarly, penetration of its body by the fungus cannot be the cause, for the eelworm appears to be, to all intents and purposes, dead before penetration begins. It is hardly likely that the animal dies from exhaustion following its desperate struggles for freedom, for eelworms are notoriously tough. Starvation can be ruled out, as the interval between capture and death is so short. We are left with the suggestion that the fungus produces a toxin which is fatal to eelworms; there is no evidence for this, but it must remain the most probable suggestion, at any rate for the time being.

As soon as a captured eelworm is dead or moribund (it is hard to say which) its integument is penetrated by a lateral branch from the fungul hypha, and this gives rise to a number of hyphae which spread throughout the carcass of the eelworm and absorb its contents. These are known as trophic hyphae. When the body contents of the prey have been completely absorbed the protoplasm of the trophic hyphae passes back into the parent mycelium, leaving the shrivelled integument of the eelworm, full of empty trophic hyphae, attached to the mycelium of the fungus.

The non-mycelial Zoopagales operate in a different way. In the genus *Cochlonema,* for example, the prey consists of amoebae or other small Protozoa. The conidia of the fungi, which are small and usually more or less cigar-shaped, are produced in considerable quantities. A conidium is picked up by one of the animals, either by adhesion or by ingestion with its food, and it germinates to produce a coiled structure inside the animal. This structure, whitch is called a thallus, grows until it practically fills the body of the host, which becomes increasingly sluggish as its body contents are absorbed by the fungus. Eventually the host dies and the fungus enters upon its reproductive stage (Fig. 27).

Asexual reproduction in the Zoopagales is by means of conidia, not sporangiospores as in the Mucorales, and the different genera are distinguished from one another by the way in which the conidia are carried. In *Zoopage* and *Cochlonema,* for instance, the conidia are usually cigar-shaped and formed in long chains. In *Stylopage* the conidia are ovoid or spindle-shaped and carried either singly or in groups on long conidio-phores that stand erect, while in *Acaulopage* they vary in shape and are formed on short projections from the hyphae. The conidia of *Acaulopage* often bear empty appendages at their ends.

Sexual reproduction in the Zoopagales follows the general pattern described in the previous chapter for the Mucorales, two gametangia fusing with one another and a zygospore being formed either at the point of fusion or on the end of a small

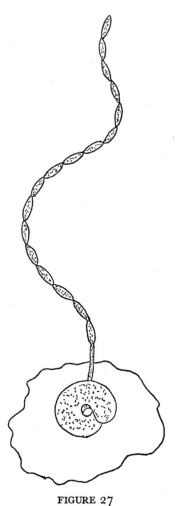

FIGURE 27

A dead amoeba containing a thallus of Cochlonema verrucosum. *The fungus has produced a fertile hypha bearing a
chain of conidia. Greatly magnified.*

branch. Many of the Zoopagales are not known to reproduce
sexually. The zygospores are small and easily recognised by the
hemispherical warts covering their surfaces.

The Zoopagales are believed to be obligate predators, since,
except for a few species that are not predacious, they have

F

never been grown without their appropriate hosts. I have many times picked conidia of *Stylopage grandis* off their conidiophores with a sterile needle and transferred them to Petri dishes containing sterile agar culture medium in the hope of growing them in pure culture, but uniformly without success. On the other hand, the conidia of many species germinate readily in mixed cultures where their hosts are present. This suggests that some substance given off by the host animal must be present if the spores are to germinate, though the existence or otherwise of such a substance has not been demonstrated.

Although they lurked unsuspected until comparatively recently the Zoopagales are, in fact, extremely common in the soil and in rotting vegetation. It may seem strange that nobody found them before, but it must be remembered that they are extremely small. Also, they do not make their presence obvious except in the presence of the small animals on which they prey. Most mycologists avoid allowing animals to get into their cultures, so that it is really not so surprising that the Zoopagales took a lot of finding.

THE EELWORM-TRAPPING HYPHOMYCETES

The other large group of predacious fungi belongs to the class Hyphomycetes of the Deuteromycotina. These fungi are in the main trappers of eelworms, though one or two species attack humbler prey such as amoebae or rotifers.

The predacious Hyphomycetes have been known for a long time, though in many instances their predacious habit was unsuspected since they were not caught *flagrante delicto*. *Arthrobotrys superba,* for instance, was first described by Corda in 1839, but it was not known to be an eelworm-trapper until 1937, nearly a century later. Again, *A. oligospora,* the commonest of all the predacious fungi, was first described by Fresenius in 1852, but although the hyphal networks with which it captures eelworms were observed by Woronin in 1870, their function went unsuspected until 1888, when it was first described by Zopf in a paper that is now one of the classics of mycology.

Even when the existence of the predacious habit became known through Zopf's paper, the predacious Hyphomycetes were regarded as rare mycological curiosities and attracted little attention. It was not until Drechsler, the discoverer of the Zoopagales, published a long paper in 1937, in which he described many new species, that mycologists realised that the predacious Hyphomycetes were, in fact, common fungi.

The predacious Hyphomycetes, unlike the Zoopagales, are provided with special organs for capturing eelworms. These are of several different kinds. In *Arthrobotrys oligospora* the hyphae which, as in all the Deuteromycotina, are septate, have at intervals collections of loops which form networks in three dimensions. The loops are adhesive, and any eelworm that accidentally touches them is held firmly by a sticky fluid which can be seen quite clearly under the microscope (Plate 13). Capture is followed by the usual frenzied struggles to get away, but the eelworm is held securely, and after about an hour the animal stops struggling and is, as far as one can judge, dead. As with the eelworm-capturing members of the Zoopagales the cause of death is uncertain, but there is some evidence that a toxin is secreted by the fungus and that this kills the eelworm.

When the eelworm is dead or moribund a fine process grows out of the hyphal network and penetrates the integument of the animal. This swells within the body of the eelworm to form a spherical infection bulb from which trophic hyphae grow out and consume the body contents of the animal. Within twenty-four hours there is nothing left but the integument of the eelworm, filled with empty trophic hyphae the protoplasm of which has passed back into the parent mycelium.

Instead of the systems of networks, some predacious Hyphomycetes possess short adhesive branches which serve the same purpose; *Monacrosporium cionopagum* and *Dactylella lobata* are examples of predacious Hyphomycetes with this kind of trap. Others, such as *M. ellipsosporum*, bear short lateral branches on the ends of which are small sticky knobs. The events following the capture of an eelworm by adhesion to the branches or knobs are essentially similar to what happens when an eelworm is caught by the adhesion networks of *Arthrobotrys oligospora*.

Some of the predacious Hyphomycetes rely on mechanical devices for the capture of eelworms. *Dactylaria candida,* for instance, produces non-constricting rings. These are small rings, with openings a little smaller than the diameter of an eelworm's body. An eelworm that accidentally pushes its head into such a ring lacks the sense to withdraw and tries to bullock its way through. As a result, it gets stuck in the ring. Capture of an eelworm is followed by the intrusion of trophic hyphae into its body in the usual way.

The most sophisticated trapping mechanism produced by the predacious Hyphomycetes is the constricting ring. This is made up, like the non-constricting ring, of three curved cells, and is attached to the mycelium of the fungus by a short, stout stalk. The three cells of the constricting ring are sensitive to touch on their inner sides, and when an eelworm intrudes its head into the ring the friction of its body stimulates the ring cells to swell suddenly inwards to about three times their former volume, almost completely occluding the opening of the ring. The time taken for the swelling of the cells to take place is no more than one-tenth of a second. The body of the captured eelworm is deeply constricted by the grip of the ring cells, so that escape is impossible (Plate 14). The rings are situated at such a distance apart that an eelworm caught near its head may, in its struggling, flick its tail into another ring and so become doubly held.

The predacious Hyphomycetes reproduce asexually by means of conidia which are borne aloft on the ends of erect conidiophores. The conidia are large, and always of more than one cell.

Unlike the Zoopagales, the predacious Hyphomycetes are easily grown in pure culture without their prey; when isolated in this way they grow in the same way as any other mould, without producing their characteristic eelworm traps. If eelworms are added to a pure culture of a predacious Hyphomycetes traps are formed and the capture of eelworms begins. Even sterile filtered water in which eelworms have lived is sufficient to initiate trap formation, so it is clear that some diffusible substance is produced by eelworms which stimulates the formation of the traps. This substance has been given the name 'nemin'.

The ease with which the predacious Hyphomycetes can be grown in pure culture has led various workers in different parts of the world to try to devise a method of using them for the biological control of plant parasitic eelworms. Some experiments have met with a limited amount of success, but a great deal of work still remains to be done before their efficacy in controlling eelworms can be fully evaluated.

As in the Zoopagales, some Hyphomycetes are internally parasitic in eelworms, examples being *Harposporium* and *Acrostalagmus*. Here the eelworm is infected by a germ hypha from a minute conidium which sticks to its body or, in a few instances, is swallowed by the eelworm. The mycelium of the fungus grows entirely inside the body of the eelworm, only the fertile hyphae emerging to bear the crop of conidia which will infect other eelworms.

These parasitic Hyphomycetes were for a long time thought to be obligate parasites, but some of them are now being grown in pure culture without their hosts.

The eelworm-attacking Hyphomycetes, both predacious and parasitic, are extremely common : almost as common as eelworms, which are virtually ubiquitous. Anywhere the eelworms go the fungi follow them; in the soil, in leaf mould, in dung, in moss cushions (where eelworms abound) and in any kind of rotting vegetation you are almost certain to find them.

ENTOMOGENOUS FUNGI

The entomogenous fungi are fungi that attack insects; *Cordyceps militaris* has already been mentioned in Chapter Eight. An interesting order of entomogenous fungi is the Entomophthorales which belongs, like the Mucorales and the Zoopagales, to the Zygomycotina.

A common member of the Entomophthorales is *Entomophthora muscae,* a fungus which attacks the house fly (*Musca domestica*). Sometimes, particularly as autumn approaches, one may see a dead house fly stuck to a window pane, with what appears to be a 'halo' on the surrounding glass. This fly has almost certainly been killed by *Entomophthora muscae*, the halo on the glass being formed by thousands of conidia of the fungus which have been discharged from conidiophores emerging from the body of the insect. The disease caused by *E. muscae* is sometimes known as fly cholera.

A house fly is infected by a conidium lodged on its body, a germ tube from which penetrates the cuticle of the fly and gives

rise to a mycelium within its body. The mycelium consists at
first of septate hyphae (the Entomophthorales are unique among
the lower fungi in having regularly septate hyphae), but later
the mycelium breaks up at the septa to form quantities of oblong
'hyphal bodies' which may increase in number either by cell
division or by budding. Eventually the body of the insect may
become filled with hyphal bodies. As can be imagined, the fly
becomes more and more sluggish in its movements, until finally
it settles on a window pane or elsewhere and dies.

After the death of the host the hyphal bodies produce conidia
on the ends of short, thick conidiophores which burst through
the cuticle of the fly in large numbers. There is no doubt that
the conidia are, in fact, modified sporangioles, for if a conidium
is placed in water the single spore can be seen, separated from
the wall of the sporangiole.

The conidia of *Entomophthora muscae* are covered with a
sticky substance, causing them to adhere to any surface with
which they come into contact. The conidia are shot off the
conidiophores with some violence; if a spore should land on a
house fly it germinates and sets up an infection, starting the life
cycle over again. Most of the spores are not so lucky. Should a
spore land on some inanimate object, however, it still has a
second chance to find an insect host, for on germination a
conidiophore is produced directly, bearing a spore which is
slightly smaller than the first one. This spore is shot off the
conidiophore in the same way as the first one and, if it misses its
target, a third, still smaller spore may be produced. This may
be repeated several times, the spores becoming gradually smaller
each time.

Entomophthora muscae is not known to reproduce sexually,
but it does produce spores called azygospores, which are thought
to be a rudiment of a lost sexual process. An azyospore is a
thick walled spore, resembling the zygospores found in the
Mucorales, formed at the end of a short branch arising from a
hyphal body. In some other species of *Entomophthora,* such as
E. fresenii and *E. fumosa,* there is a normal sexual process in
which zygospores are formed as a result of copulation between
two hyphal bodies.

Fungi belonging to the Entomophthorales do not confine their
attentions to house flies. *Entomophthora grylli,* for instance,

attacks grasshoppers in some parts of the United States. The grasshoppers often climb up the stems of grasses before dying, and their bodies can be seen ringed with belts of conidiophores which push their way out between the segments of the abdomen.

Species of *Massospora*, a genus allied to *Entomophthora* attack the seventeen-year locust, *Tibicina septidecim*. In this genus the spores are produced inside the host, and are disseminated by the insects sloughing off the hinder segments of its abdomen as the disease progresses. Infected locusts can often be seen crawling about with only one or two abdominal segments left behind the head and thorax.

13 The lower fungi

We come now to the last and most primitive subdivision of the fungi, the Mastigomycotina. This is an extremely heterogeneous group comprising three classes which do not appear to be very closely related to one another. One thing they do have in common, however, and that is their motile asexual spores.

The asexual spores of the Mastigomycotina are called zoöspores because they are provided with either one or two minute protoplasmic 'tails', or flagella (Latin *flagellum*, a whip), by the waving movements of which they are able to swim. The word zoöspore comes from the resemblance of these swimming spores to tiny animals. Only a few species of the many hundreds contained in the Mastigomycotina have non-motile spores.

The Mastigomycotina are divided into three classes according to how the flagella are arranged in the zoöspores. In the Oömycetes the zoöspores have two flagella which may be carried either apically or laterally. The Hyphochytridiomycetes are a very small group in which there is one flagellum mounted at the anterior (front) end of the zoöspore, while the Chytridiomycetes have zoöspores with their single flagellum at the posterior (rear) end.

THE DOWNY MILDEWS

The downy mildews belong to the family Peronosporaceae of the class Oömycetes. They are all plant parasites, and they get their name from the appearance of the host plants, which are covered with a white, fluffy down (compare with the powdery mildews, Chapter Nine). The fluffy appearance is due to the production by the fungus of numerous sporangiophores bearing zoösporangia (Plate 15).

The downy mildews cause a number of important plant diseases, perhaps the most important, and certainly the most famous of which is the downy mildew of the grape vine which nearly destroyed the wine industry in France during the second half of the nineteenth century. The story of this is interesting, not only as an example of how a parasitic fungus can threaten the economy of a whole nation, but also because the French outbreak of downy mildew of the vine marks the first time in history that a fungicide spray was used on a large scale to combat a plant disease.

Downy mildew of the vine is caused by *Plasmopara viticola,* a fungus that was originally a native of America. Around the middle of the eighteenth century the French vineyards were all but destroyed by an insect pest called *Phylloxera,* an aphid that attacked the roots, also a native of America, where it parasitises the American grape, *Vitis rupestris.* The behaviour of the insect in America is different from its behaviour in Europe; it attacks only the foliage of *Vitis rupestris,* whereas with the European grape, *V. viticola,* it leaves the foliage alone and goes for the roots. During the great European epizoötic of the nineteenth century somebody had the bright idea of grafting *Vitis viticola* stems on to *V. rupestris* roots; in this way, it was hoped to produce a plant which *Phylloxera* could not attack. This was done; at great cost and even greater effort all the grape vines in Europe were grubbed up and replaced by the new *rupestrisviticola* grafts. The expense and effort were justified: *Phylloxera,* finding neither the foliage nor the roots of the new plants to its taste, retired to high dudgeon, and the French wine crop was saved.

Unfortunately, the grafting experiment brought its penalty in the form of *Plasmopara viticola.* Sometime or another – nobody knows when or how – some American grapes brought into France for grafting carried infection with them. In America *P. viticola* is not a serious pest, for it has been co-existing with the American grapes for many thousands of years, and the grapes have in that time evolved a certain degree of resistance to the fungus. The European grape, however, had no such resistance, and when *P. viticola* was introduced into Europe it ran through the vineyards like a fox through a chicken run. The French wine industry, still reeling from the attack of *Phylloxera,* seemed to be doomed.

Just when things looked really hopeless help came from a

great botanist called Millardet, Professor of Botany at Bordeaux. Millardet chanced to be walking through a vineyard in the Médoc district and he noticed that the vines immediately on the verge of the pathway he was using appeared to have escaped attack by downy mildew, although the rest of the vines were ravaged. He also noticed that the healthy vines appeared to have been sprayed with something resembling verdigris. Millardet was interested. He made inquiries, and was told by the owner of the vineyard that people using the pathway on a hot summer's day found the grapes irresistible, so the vines near the path had been sprayed with a mixture of copper sulphate and lime in the hope of discouraging theft.

This was enough for Millardet. He felt sure that the copper spray was doing something that prevented *Plasmopara viticola* from attacking the vines that had been sprayed, and he went straight back to his laboratory to experiment. From his work was born the famous Bordeaux mixture, a basic copper hydroxide formed by mixing copper sulphate with lime, which is still regarded as one of the most effective fungicides.

The discovery of Bordeaux mixture undoubtedly saved the vine, but there still remained the work of spraying it on to the growing crop, which was a labour of gangantuan magnitude; remember, this was the *first* time a crop had ever been sprayed, so there were no mobile sprayers such as we use today. The wretched labourers walked through the vineyards with buckets of Bordeaux mixture which they splashed on to the vines with whisks made of heather. There were many aching French backs before the end of the spraying season. Later various machines were constructed for doing the job mechanically, some of them more noted for their ingenuity than for their convenience. There was, for instance, the spraying machine that worked by a pair of bellows incorporated into the boots of the operator!

Plasmopara viticola attacks the leaves of the vine, forming a mycelium of non-septate hyphae that run between the leaf cells, drawing nourishment from the host by means of haustoria which are inserted into the cells. After a time hyphae grow out through the stomata (breathing pores) of the leaf; these are the sporangiophores. They usually branch two or three times, a more or less globose zoösporangium being formed at the tip of each branch (Fig. 28).

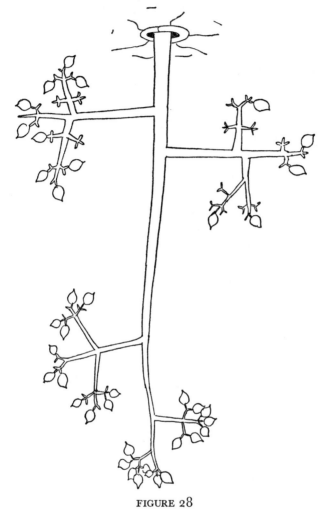

FIGURE 28

A sporangiophore of Plasmopara viticola *emerging from a stoma on the surface of the leaf of a grape vine. The small pear-shaped bodies are sporangia. Greatly magnified.*

The zoösporangia are abstricted from the ends of the sporangiophores and blown away by the wind; they function, in fact, in the same way as conidia. When they fall on a vine leaf they liberate multitudes of zoöspores, which are tiny bean-shaped objects, each with a pair of lateral flagella; all the Peronosporales have zoöspores of this type. The zoöspores are able to swim in the film of moisture that covers the leaf surface, and they infect more vines with the fungus. By successive crops of zoöspores the disease is rapidly spread through a vineyard.

Towards the end of the growing season, as the leaves of the vine begin to die, it is necessary for the fungus to enter its overwintering stage, for the life of the zoöspores is evanescent and the mycelium needs a vine leaf to house it. It is then that sexual reproduction takes place. The female sex organ is a spherical structure at the end of a hypha; it is called the oögonium, and within it is formed the female sex cell or egg cell. The male organ, or antheridium, is filamentous, resembling a normal hyphal branch. The tip of the antheridium attaches itself to the oögonium, and a nucleus passes from it into the oögonium and fuses with the nucleus of the egg cell. When the cell has thus been fertilized it surrounds itself with a thick wall and remains dormant in the soil until the following spring when the vines are again in leaf.

This type of sexual reproduction where an already differentiated egg cell is fertilised by a male gamete of some kind or another is called oögamy. The thick-walled spore that results from it is called an oöspore.

POTATO BLIGHT

Late blight of potatoes is an exceedingly common disease, caused by a fungus called *Phytophthora infestans*. If you visit almost any allotment in mid-September you will notice that the leaves of the potato plants are turning brown and that the plants are beginning to die off. If the allotment holder is present at the time he will point out that his potatoes are 'ripening nicely'. Nothing could be farther from the truth. In fact, the potato plants are being killed by *Phytophthora,* which is the normal fate of a potato crop in temperate latitudes, unless it has been sprayed with Bordeaux mixture or some other fungicide. If sprayed, the potato plants would have remained green for another month, and the potao crop would have been that much heavier.

Phytophthora infestans is another member of the Peronosporales (family Pythiaceae), with a life history essentially similar to that of *Plasmopara viticola*. It usually begins to attack the potato crop in earnest after the first rains in July, the moisture on the foliage helping the spread of the disease by zoöspores. The attack is always worse in a wet season, and in a very wet summer it can be catastrophic. It attacks the foliage of the potato plants while they are growing, and at the end of the season, if the spores should get into the potato clamps, it causes rotting of the tubers in the clamp. Many a farmer has opened his potato clamp in January expecting to find healthy potatoes and has found instead a stinking, rotting mess.

It was *Phytophthora infestans* that caused the great Irish famine during the middle of the eighteenth century, when potato blight hit the Irish potato crop with devastating severity. At that time the Irish peasantry lived on potatoes and whiskey and, as the whiskey was distilled at home from fermented potatoes, they were deprived at one stroke of both food and solace. It is hard to get reliable figures of the effects of such a disaster, but it was certainly the worst catastrophe to strike the British Isles since the Black Death, and it is estimated that upwards of a million Irish died of starvation in the years 1845-1860, and that another million and a half emigrated to America. The large Irish element in the population of the United States is closely related with *Phytophthora infestans*.

THE WATER MOULDS

The water moulds, or Saprolegniaceae, are, as their name implies, aquatic fungi, using the term 'aquatic' in its widest sense as many species of Saprolegniales are inhabitants of damp soil rather than ditches and ponds. One must remember that these fungi are so small that to them the film of moisture covering every particle of wet soil is shallow water.

Most of the Saprolegniaceae are saprobes which occupy a position in aquatic environments similar to that of the moulds belonging to the Mucorales on land. A few are parasitic; *Saprolegnia parasitica*, for instance, causes diseases of fish and their eggs, and some species of *Aphanomyces* attack higher plants. Most of the Saprolegniaceae, however, grow on dead organic matter in water or damp soil: the decaying bodies of insects or their larvae, decaying fruits and seeds, and so on. If you should find the dead body of a fly floating in a stream and covered with a mass of rather coarse fungal mycelium you can be fairly sure that here you have one of the Saprolegniaceae, for they are noted for the large diameter of their non-septate hyphae.

The Saprolegniaceae reproduce asexually by means of zoöspores which, as in all the Oömycetes, are biflagellate. Most of the Saprolegniaceae, however, have two kinds of zoöspores, one kind with apical flagella and the other with lateral flagella. The zoösporangia are cylindrical or club-shaped (Fig. 29 A, B). In *Saprolegnia* the zoöspores are at first pear-shaped with the flagella inserted at the pointed end; these are released from an opening at the apex of the zoösporangium and they swim away. These are called primary zoöspores (Fig. 29 C). After a time, however, the primary zoöspores stop swimming, withdraw their flagella, and surround themselves with a wall, forming a rounded structure called a cyst. Presently the cyst opens again and the zoöspore emerges, this time with its flagella inserted at one side. These zoöspores are called secondary zoöspores (Fig. 29 D). They swim around for a time and then, coming to rest on a piece of organic matter of some kind, each zoöspore puts out a germ hypha and produces a new mycelium.

The encystment of the zoöspore and its emergence with altered flagellation is called diplanetism. Typical diplanetism is shown by *Saprolegnia, Isoachlya, Leptolegnia,* and *Leptolegniella.* There is a tendency in the Saprolegniaceae for diplanetism to be shortened, one type of zoöspore being either of short life or suppressed altogether. In *Achlya* the primary zoöspores encyst as soon as they leave the sporangium, the cysts being gathered in a cluster round the sporangial mouth (Fig. 29 B); in due course the secondary zoöspores emerge from the cysts and swim away. In *Dictyuchus* the primary zoöspores are never liberated; they encyst inside the sporangium and secondary zoöspores are later set free, each through a separate opening in the zoösporangial wall. In *Geolegnia* and *Aplanes* both primary and secondary zoöspores are suppressed, non-motile spores (aplanospores) being formed in the sporangium and germinating either in the sporangium or after the sporangium has disintegrated. It is possible that the sporangiospores found in the

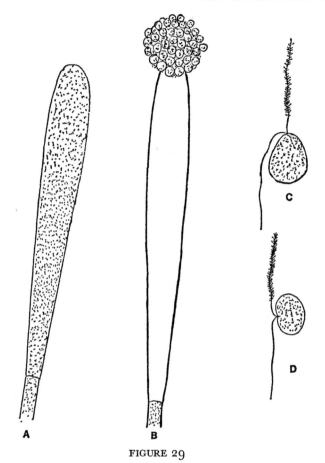

FIGURE 29

A, zoösporangium of Saprolegnia. *B, empty zoösporangium of* Achlya, *with zoöspores encysted round the mouth. C, primary and D, secondary zoöspores of* Saprolegnia.

Zygomycotina may have evolved from some arrangement such as this.

It is not always the primary zoöspore whose life is shortened or suppressed. In *Pythiopsis* the primary zoöspores are the only ones to be formed; when they have finished their active stage they germinate and form new mycelia.

Sexual reproduction in the Saprolegniaceae follows the

general pattern that has already been described for the Peronosporales, the most obvious difference being that in most of the Saprolegniaceae the oögonium contains several egg cells, all of which are often fertilised by nuclei from one antheridium, though there may be more than one. A less obvious difference which is important to mycologists is that in the Peronosporales the oögonium contains some residual protoplasm which is not included in the egg cell, whereas in the Saprolegniaceae all the protoplasm in the oögonium goes into the formation of the egg cells.

THE MOST PRIMITIVE FUNGI

The most primitive known fungi are to be found in the class Chytridiomycetes, and particularly in the order Chytridiales. This order contains species that inhabit water and soil; some of them are saprobes and others are parasitic on algae or watermoulds, or, in the case of soil-inhabiting species, on higher plants. Few are of any economic importance but one or two attack crop plants, outstanding among which is *Synchytrium endobioticum*, which causes wart disease in potatoes. *Urophlyctis alfalfae*, the cause of crown wart of lucerne, may produce serious crop losses at times.

The Chytridiales are all extremely small. They are all thalloid (that is, they have no mycelium of hyphae), and the simplest parasitic species consist of small, single cells contained within the cells of their hosts. Others have their single cells perched on the outside of their host, or on the surface of decaying organic matter if they are saprobes, with a few fine strands, called rhizoids, penetrating the substratum. A few, known as polycentric chytrids, have several centres of activity – one cannot call them cells – connected with one another by fine strands of protoplasm known as the rhizomycelium.

The Chytridiales, like all the Chytridiomycetes, reproduce asexually by means of zoöspores, each of which has a single posterior flagellum. Often the whole of the vegetative body of the fungus becomes a zoösporangium when mature, its protoplasm splitting up to form zoöspores. Their sexual reproduction is so varied that it is impossible to generalise about it.

The Chytridiales are so varied in their structure and their

life histories that no one example can be said to be typical. I am going to describe *Olpidium viciae*, not as a typical example of the Chytridiales, but to show the kind of organisation that is found in an extremely primitive fungus.

Olpidium viciae is a parasite on the stems and leaves of *Vicia unijuga*, a plant of the vetch family (Papilionaceae). Zoöspores of the fungus swim around on the surface of the host plant during wet weather and then encyst. From a cyst the protoplasm emerges, dissolves a small hole in the host cell wall, presumably by enzyme action, and enters the host cell, leaving the empty cyst behind. Once inside a cell of the host the minute blob of protoplasm attaches itself to the nucleus of the cell, surrounds itself with a membrane, and grows until it nearly fills the cell. The thallus of the parasite finally divide up into numerous zoöspores, each with a single posterior flagellum, which are liberated through one or more exit tubes which grow out of the sporangium and pierce the cell wall of the host. After swimming around for a time the zoöspores infect other host cells.

The sexual reproduction of *Olpidium viciae* is as simple as the rest of its life history. Two zoöspores fuse together, forming a biflagellate cell which invades a host cell in the same way as a zoöspore. It then grows into a thick-walled sporangium which can remain dormant through the winter if necessary, its contents dividing up in the spring to form zoöspores.

14 Collecting fungi

On the whole, the larger fungi are not good material for the collector, for they do not lend themselves to preservation. Some of the agarics dry quite well, though shrinkage is considerable and their colours alter, but most become valueless after a few days, if not sooner, either through decomposition or the attacks of mites, to which most fungi are extremely susceptible. Fungi may be preserved by freeze-drying, but this needs expensive and bulky equipment and is hardly within the compass of the amateur mycologist. The best place to examine fungi is in the field at the time of collection; if time does not permit this, they should be brought home, treating them with the tenderest care because they are very easily damaged, and examined as soon as possible.

There is only one kind of container suitable for carrying fungi: a shallow, open shopping basket. This should be lined with moss, grass, soft cloth, or at a pinch paper, to protect the specimens from damage if they are jolted about. Small specimens should be packed in separate containers, such as old tobacco tins padded with moss or some soft material. Some mycologists carry tobacco tins with a piece of cork mat on the bottom of the tin to which small specimens can be pinned; this is all right as long as the pins stay in place, but they seldom do.

If you wish to be really thorough, wrap each specimen in a piece of newspaper before placing it in the basket. This will prevent spores from one specimen from cascading over the specimens beneath, affecting their colour.

Little other apparatus will be needed for collecting fungi. A pocket notebook and a lead pencil will be needed to record where each specimen was found and any other relevant details. Do not rely on a fountain pen, ballpoint, or, worst of all, an indelible pencil, for the paper may get wet, causing the ink to run. A pocket knife with a large blade is essential, not only for sharpening the pencil, but also for prizing agarics out of the ground, or cutting the bark or wood on which bracket fungi or cup fungi may be growing. Don't forget that an agaric must always be prized out of the ground from beneath in order that the volva, if any, may be removed with the fungus.

A good hand lens is quite a useful adjunct. Get one magnifying ten diameters, consisting of a single lens which swings out of a cover, not one of the kind where two or three lenses are mounted separately with a common cover; they are hopeless. If you want to be kind to yourself, and can afford it, get the type of lens known as 'aplanatic'; such lenses are a little expensive, but worth the money for the much better image they provide.

One other thing you will need to get: a good book to enable you to identify the specimens you have collected. I particularly recommend *Wayside and Woodland Fungi* by W. P. K. Findlay, published by Warne. It describes all the fungi you are likely to find, and the coloured illustrations, many of them by the legendary Beatrix Potter of Peter Rabbit fame, are really superb. Apart from its value for the identification of specimens, the book is by an eminent mycologist with a real feeling for fungi (the two things do not necessarily go together, unfortunately), and it contains a mass of interesting information which makes it much more than a field guide.

Fungi are extremely difficult to identify, so do not get too depressed if at first you have many failures. With a very little practice you should be able to give the genus correctly to most of your specimens, but the species will all too often elude you. Life is made more difficult for the mycologist by the fact that many fungi, especially agarics, change so rapidly once they have been gathered, especially in colour, and colour is most important in distinguishing between species of the same genus. Some agarics become unidentifiable in a few hours. There is not much you can do about this, except to carry a book with you and identify on the spot whenever possible. In any case, always make exact notes of the appearance, and particularly the colour, of a specimen before it is gathered.

Do not forget to make a note of where each specimen was growing, and of what it was growing on. This will be a valuable, and in most cases an essential, clue to its later identification. Sometimes the habitat will be the only clue needed to identify a fungus with virtual certainty; a small species of *Hydnum* growing on pine cones, for instance, can be diagnosed as *Hydnum auriscalpium* almost without looking at it.

Attempts to preserve fungi indefinitely, as I have said, usually do not pay, though there are exceptions. Woody fungi will often keep quite well if dried and placed in boxes, a little naphthalene being added to discourage the small beetles that are often a nuisance to the collector.

A useful method of preserving fleshy agarics is to cut a thin slice (section) from the middle of the cap down to the bottom of the stalk, and to dry this in a press as one would do in preserving a specimen of a flowering plant. The section is placed between several thicknesses of blotting paper (old newspaper can be used at a pinch) and left under a weight. The drying paper must be changed, at first daily, and then every two or three days until the specimen is dry, when it can be stuck with narrow strips of adhesive paper on to a sheet of cartridge paper and placed in the herbarium. The name of the fungus, the habitat, the date of the collection, and any other relevant details should be written on the right hand bottom corner of the herbarium sheet. Herbarium sheets of the correct size (16 x 11 inches) may be obtained from any dealer in biological supplies, and also printed herbarium labels on which to write the details, but such nicety is not really necessary.

It is also useful to preserve spore prints of agarics. The manner of obtaining them is described in Chapter Two, and they can be made more or less permanent by blowing over them a little of the fixative that artists use for charcoal drawings.

If you have a microscope you may wish to collect microscopic fungi such as moulds. A discussion of methods used by microscopists in handling fungi cannot be undertaken here, but a few brief hints may be of value to those microscopists who have not handled fungi before.

The best fixative and preservative for microscopic fungi is formalin-acetic, made up as follows:

Glacical acetic acid	5 per cent by volume
Commercial formalin	10 per cent by volume
Water	85 per cent by volume

Specimens can be left in this fluid for years without coming to any harm.

Fungi should, wherever possible, be examined under the microscope living and unstained; in this way their true structure can be observed. If permanent preparations are desired the best medium for mounting and staining is lactophenol cotton blue. Lactophenol is made up as follows:

Phenol crystals	10 g.
Lactic acid, sp.gr. 1.21	10 g.
Glycerol	20 g.
Distilled water	10 g.

Sufficient powdered cotton blue is added to the lactophenol to make it the colour of ink.

To mount in lactophenol cotton blue, a minute portion of the fungus is placed in a drop of the lactophenol cotton blue on a microscope slide and teased up thoroughly with a pair of mounted needles. The slide is warmed over the smallest possible flame obtainable with a Bunsen burner, or a spirit lamp, until steam begins to rise. Do not boil. Allow the slide to cool for a quarter of a minute and then carefully apply a cover slip. I emphasise 'carefully' because it is extremely difficult – I am tempted to say impossible – to cover a lactophenol mount without including some air bubbles, which have to be removed by subsequently heating the slide.

Lactophenol cotton blue kills, fixes, and stains in one operation. As most micro-fungi are extremely delicate, this is just as well, as the more handling that can be avoided the better. The contents of hyphae are stained blue, while cell walls are not stained. To make the mount permanent, seal the edges of the cover glass with red nail varnish, giving two coats. The more purplish and vulgar-looking the nail varnish, the better it seems to stick; for this I cannot account. One word of warning: the amount of lactophenol used must be judged exactly to fill the cover glass and no more, for if so much as a micro-drop escapes round the edge the seal will be useless. Incidentally, lactophenol mounts will keep unsealed for weeks without coming to any harm, since lactophenol evaporates very slowly. Some authorities prefer to keep their mounts unsealed for a week or so before sealing.

Micro-fungi abound everywhere, awaiting to be collected and looked at. For the amateur working at home I particularly recommend the aquatic fungi. Collect about a quart of water from a stagnant pond, bring it home, and 'bait' it for fungi with rose hips, boiled hemp or cress seeds, dead house flies, or anything else you like. After about a week, pick the bait out of the water and examine it for fungi; you will almost certainly find water moulds of one kind or another (remember that there are far more water moulds than I had space to describe in this book).

Collect also samples of the scum to be found floating on the surface of any pond. Filaments of algae such as *Spirogyra* may well show chytrids or other parasites. Many predacious fungi occur in water, attacking amoebae, rotifers, and eelworms. You might well see some of them.

If a specimen of fungus bait from your water has a fungal mycelium growing on it which is not yet forming sporangia it can be put back into the water and allowed to develop further. Alternatively, it can be put in water in a small container apart from the main batch.

Dung is a good source of fungi of many kinds. If a sizeable lump of horse or cow dung is placed on a plate or board and covered with a large inverted plastic bag a regular succession of fungi will grow out of it. First will come the Mucorales, including *Pilobolus* with its interesting cap-throwing activities. A little later these may be seen to be attacked by parasitic Mucorales such as *Piptocephalis* and *Chaetocladium*. After the Mucorales have had their day they will be followed by small Ascomycotina such as *Sordaria* and *Ascobolus*. Finally, small agarics will appear, particularly tiny species of *Coprinus*. A good piece of dung can keep a mycologist happy for weeks.

If, as I hope, you develop a genuine enthusiasm for fungi and wish to study them in company with others, there is the British Mycological Society to join. This caters for all mycologists, professional and amateur, and even the raw beginner will be welcome at the meetings if he has a genuine interest in fungi. Fungus forays are held in the autumn at places near London, as well as at Provincial centres, where fungi are collected and identified by experts. Paper reading meetings are also held periodically in London and elsewhere, and the Society's *Transactions* alone are well worth the modest subscription.

GLOSSARY

Terms of little importance that have been mentioned only once in the text have, in general, been omitted from the glossary.

Adnate. The condition, of the gills of a toadstool, where the gills are fused with the stem for the whole of their width.

Adnexed. The condition, of the gills of a toadstool, where the gills are fused with the stem for only part of their width.

Aecidiospore. The spores formed in an aecidium of a rust fungus.

Aecidium. A cup-shaped structure produced by a rust fungus and containing chains of aecidiospores.

Anaerobic respiration. Any form of respiratory process in which air is not needed.

Anisogamy. Sexual reproduction in which the gametes that fuse differ from one another in size but not in general form.

Antheridium. The male sex organ of a fungus.

Aplanospore. A non-motile spore.

Apothecium. A fruit body formed by an ascomycete in which the asci are contained in a cup-shaped structure, or a modified structure arising from an inverted cup.

Arbuscle. A bunch of fine hyphal branches produced by a fungus causing vesicular-arbuscular mycorrhiza.

Ascogonium. The female sex organ of an ascomycete.

Ascogenous hyphae. The special hyphae on which the asci of an ascomycete are formed.

Ascomycete. A fungus belonging to the Ascomycotina.

Ascospore. A spore formed in an ascus.

Ascus. A special type of sporangium formed by an ascomycete, containing (usually) a limited number of ascospores.

Basidiocarp. The fruit body of basidiomycete.

Basidiomycete. A fungus belonging to the Basidiomycotina.

Basidiospore. A spore formed on a basidium.

Basidium. A cell in which nuclear fusion is followed by meiosis and exogenous spore formation. Basidia are found in the Basidiomycotina.

Bordeaux mixture. A fungicide made by mixing copper sulphate with lime.

Chromosome. One of the usually rod-shaped bodies found in the nuclei of cells, on which the hereditary information is carried.

Cleistothecium. A fruit body found in certain of the Ascomycotina where the asci are contained in a closed structure without an ostiole for the release of the ascospores.

Columella. The dome-shaped septum found in the Mucorales at the tip of the sporangiophore.

Conidiophore. A fertile hypha which bears conidia.

Condium. An exogenously formed asexual spore.

Decurrent. The condition, of the gills of a toadstool, where the gills run down the stem.

Dikaryophase. The dikaryotic stage in the life history of a fungus.

Dikaryotic. The condition where the cells of a hypha each contain two nuclei of opposite mating phase.

Diplanetism. The condition where the zoöspores of a fungus encyst and re-emerge with their flagella differently inserted.

Diplobiontic. A term used of yeasts which only have a diploid vegetative stage in their life histories.

Diploid. Having the double chromosome number.

Ectotrophic mycorrhiza. A mycorrhizal association where the hyphae of the fungus are mainly superficial on the outside of the root.

Endotrophic mycorrhiza. A mycorrhizal association where the hyphae of the fungus penetrate extensively into the root of the host.

Enzyme. An organic catalyst.

Facultative saprobe. A parasitic fungus which can also live saprobically.

Fission yeast. A yeast that reproduces by transverse fission instead of by budding.

Flagellum. A whip-like protoplasmic process which, by its waving movement, enables an organism or a spore to swim.

Free. The condition, of the gills of a toadstool, where the gills are free from the stem.

Funicle. The elastic stalk attaching a peridiole of a bird's nest fungus to the cup in which it is formed.

Gametangium. A term applied to the sex organs of the Zygo-mycotina. Also, any organ that contains gametes.

Gamete. A sex cell.

Haplobiontic. A term used of yeasts in which the whole of the vegetative stage in the life history is haploid.

Haplodiplobiontic. A term used of yeasts which have both haploid and diploid vegetative stages in their life histories.

Haploid. Having the single chromosome number.

Hartig net. The network of hyphae that spreads between the cells of the cortex of the root in an ectotrophic mycorrhiza.

Haustorium. A hyphal process or 'sucker' by means of which a parasitic fungus draws nourishment from a cell of its host.

Heterothallism. The phenomenon, found in many of the fungi, of the existence of two mating strains of the same species, usually known as the 'plus' and 'minus' strains. Sexual reproduction can only take place between mycelia of opposite strains.

Host specificity. The ability of a parasitic fungus to attack only one species of host plant, or a few related species of host plants.

Hymenium. The layer in a fruit body consisting of the actual cells that produce the spores, containing either basidia or asci as the case may be.

Hypha. One of the threads of which the mycelium of a fungus is composed.

Hyphal body. One of the cells into which the mycelium of one of the Entermophthorales fragments.

Iosgamy. Sexual reproduction in which the two fusing gametes or gametangia are similar to one another.

Karyogamy. The fusion of the two gametic nuclei in sexual reproduction.

Macrocyclic rust. A rust fungus which has one or more dikar-yotic spore types in addition to teleutospores.

Meiosis. A form of nuclear division in which the number of chromosomes in the daughter nuclei is halved.

Microcyclic rust. A rust fungus which produces no dikaryotic spore types other than teleutospores.

Mitosis. The normal form of nuclear division in which the chromosomes of the daughter nuclei resemble those of the parent nucleus in form and number.

Monokaryotic. Referring to a cell in which only one kind of nucleus is present.

Mycelium. The sum total of all the hyphae making up the body of a fungus.

Mycorrhiza. A symbiotic association between a fungus and the roots of a higher plant.

Obligate parasite. A parasitic fungus which is incapable of saprobic life.

Oögamy. A type of sexual reproduction in which an organised egg cell is fertilised by a male gamete.

Oögonium. The cell in which the egg cell is contained.

Oöspore. A spore, usually a thick-walled resting spore, formed as a result of the fertilisation of an egg cell by a male gamete.

Partial veil. The membrane that encloses the gills of an immature member of the Agaricales.

Peridiole. One of the discrete portions into which the gleba of a bird's nest fungus is broken up.

Peridium. The membrane, usually of at least two layers, that surrounds the fruit body of a gasteromycete.

Perithecium. The flask-shaped structure in which the asci of a pyrenomycete are enclosed.

Phialide. A usually flask-shaped cell specifically modified for the production of conidia.

Plasmogamy. The union of the plotoplasm of the two gametes that precedes karyogaky in sexual reproduction.

Promycelium. The short hypha that bears the basidiospores when the teleutospore of a rust or smut fungus germinates.

Pycniospore. One of the spore-like bodies formed in a pycnium.

Pycnium. A flask-shaped structure in which the pycniospores of a rust fungus are formed.

Receptive hypha. A short hypha, arising from a pycnium, with which a pycniospore may fuse.

Rhizomorph. A thread-like structure, rather like a bootlace, formed from the interwoven hyphae of a fungus.

Ring. The remains of the partial veil on the stem of an agaric.

Sclerotium. A hard resting structure formed of compacted hyphae.

Septum. A transverse partition running across a hypha.

Sinuate. The condition, of the gills of a toadstool, where the gills have a wavy margin.

Sporangiole. A small sporangium formed by one of the Zygo-mycotina.

Sporangiophore. A fertile hypha bearing a sporangium.

Sporangium. A cell in which spores are produced.

Sporophore. The fruit body of a basidiomycete.

Sterigma. A projection from a cell bearing a spore.

Stroma. A structure formed of compacted hyphae in or on which the asci are formed in many of the Ascomycotina.

Symbiosis. The state where two or more organisms inhabit the same body for mutual benefit.

Teleutospore. The spore stage in a rust or smut fungus in which the sexual fusion of nuclei takes place, and from which the promycelium bearing the basidiospores arises.

Thallus. In general, a plant body which is not divisible into stem and leaf. In a fungus, a body of indeterminate shape which is not composed of hyphae.

Trichogyne. The terminal cell in an ascogonium with which the antheridium fuses during sexual reproduction.

Universal veil. The membrane which encloses the whole fruit body of an agaric when it is immature.

Uredospore. A spore stage in the rust fungi which, in a macro-cyclic rust, precedes the formation of teleutospores.

Vesicle. A small, usually spherical swelling on the end of a hypha. A vesicle may also arise from a zoösporangium, as in the case of *Pythium.*

Vesicular-arbuscular mycorrhiza. A type of mycorrhiza charac-terised by the production of vesicles and arbuscles on the hyphae of the fungal partner.

Volva. The cup-like structure, formed from the remains of the universal veil, in which the stalk of a toadstool is set.

Zoösporangium. A sporangium in which zoöspores are formed.

Zoöspore. A spore which is provided with one or two flagella by which it has the power of movement.

Zygospore. A usually thick-walled spore formed as the result of isogamous or anisogamous sexual reproduction.

Zygote. A cell formed by the fusion of two gametes.

SELECTED BOOK-LIST

Alexopoulos, C. J. *Introductory mycology* (John Wiley, 1962).

Christensen, C. M. *The molds and man* (University of Minnesota Press, 1961).

Duddington, C. L. *The friendly fungi* (Faber and Faber, 1957).

Duddington, C. L. *Micro-organisms as allies* (Faber and Faber, 1961).

Findlay, W. P. K. *Wayside and Woodland Fungi* (Warne, 1967).

Ingold, C. T. *The biology of fungi* (Hutchinson, 1967).

Large, E. C. *The advance of the fungi* (Jonathan Cape, 1940).

Lange, M. and Hora, F. B. *Collins guide to mushrooms and toadstools* (Collins, 1963).

Ramsbottom, J. *A handbook of the larger British fungi* (British Museum, 1944).

Ramsbottom, J. *Mushrooms and toadstools* (Collins, 1953).

Robinson R. K. *Ecology of fungi* (English Unversities Press, 1967).

Rayner, M. C. and Nielson-Jones, W. *Problems in tree nutrition* (Faber and Faber, 1944).

Webster, J. *Introduction to fungi* (Cambridge University Press, 1970).

Index

Figures in *italic* represent pages where text-figures of the subject occur

Absidia, 132, 134
Acaulopage, 144
Achlya, 158, *159*
Acrostalagmus, 149
Actinomycetes, antibiotics derived from, 111–12
Agaricus arvensis, 22; *A. bisporus,* 22, 38; *A. campestris,* 11, 21, 22, 25–9, 27, 38; *A. silvaticus,* 38–9; *A. silvicolor,* 38–9; *A. xanthoderma,* 38
Aleuria aurantia, 82, 88, *89*
Amanita muscaria, 30–1, 33; *A. phalloides,* 25, 34–6; *A. rubescans,* 36; *A. virosa,* 36
antibiotics, 107–112
Aphanomyces, 157
Aplanes, 158
Armillaria mellea, 37, 59
Arthrobotrys oligospora, 146–7; *A. superba,* 146
Ascobolus, 166
ascus, formation of 84–6, *85*
Aspergillus, 37, 105, *112,* 112–14, 139; *A. niger,* 14, 113, 139; *A. ochraceus,* 112
aureomycin, 112
Auricularia auricula-judae, 51

bakers' yeast, manufacture of, 125–6
beef steak fungus, 44
bird's nest fungi, 66–7
black bulgar, *89,* 90
black rust of wheat, 13, 69–70, *71*
Blakeslea, 138
blusher, 36

Boletus badius, 41; *B. bovinus,* 55, *B. edulis, 35,* 41; *B. parasiticus,* 42; *B. satanus,* 41–2
Bordeaux mixture, 154, 156
bracket fungi, 43–9
bread, 124–5
brewing, 119–22
brown rot of apples, 11, 13
Bulgaria inquinans, 89, 90
bunts, 78–81, *80*

Calvatia gigantea, 14, 63
Cantharellus cibarius, 37
cap-thrower, 134–6, *135*
cellar fungus, 48
Cenococcum graniforme, 55
Chaetocladium, 138, 166
chanterelle, 37
Choanephora, 138
Choiromyces meandriformis, 92
citric acid, manufacture of by fermentation, 113
Clathrus cancellatus, 66
Clavaria, 50; *C. cinereus,* 50; *C. inaequalis,* 50; *C. pistillaris,* 50
Claviceps purpurea, 98–101, *100*
Clitocybe geotropa, 41
club fungi, 50
Cochlonema, 144; *C. verrucosum,* 145
Coniophora cerebella, 48
Coprinus, 29, 33, 87, 166; *C. comatus, 35,* 38
coral spot, 97–8
Cordyceps capitata, 101; *C. militaris,* 101; *C. norvegica,* 93; *C. ophioglossoides,* 101; *C. sinensis,* 101

corn smut, 79
Craterellus cornucopioides, 36–7
Cronartium ribicola, 77
crown wart of lucerne, 160
Cunninghamella, 138
cup fungi, 82–92

Dacryomyces deliquescens, 52–3
Dactylaria candida, 147–8
Dactylella lobata, 147
death cap, 25, 34–6
destroying angel, 36
Dicranophora, 137, 138
Dictyuchus, 158
Dispira, 139
downy mildews, 152–6
dry rot, 13, 45–9
dung, fungi in, 166
Dutch elm disease, 13

earth balls, 64
earth stars, 63
earth tongue, 82, 90
Elaphomyces, 101
elf cup, 88; elf cup, scarlet, 90
Endogone, 60
entomogenous fungi, 149–51
Entomophthora, 151; *E. fresenii,* 150; *E. fumosa,* 150; *E. grylli,* 150–1; *E. musci,* 149–50
ergot, 98–101, *100*
Erysiphe, 105; *E. cichoracearum,* 103; *E. polygoni,* 103
Eurotium, 113

fairy ring fungus, 39
fairy rings, 39–42
Fistulina hepatica, 44
flask fungi, 93–101
fly agaric, 30, 31, 33
formalin-acetic fixative, 164
fungus artillery, 67–8, *68*

Ganoderma applanatum, 43
Geastrum, 63; *G. triplex,* 63
Geoglossum, 82; *G. cookeianum,* 90
Geolegnia, 158
giant puffball, 14, 63
Gibberella fujikuroi, 98

gills, shape of, 32, *33*
gluconic acid, manufacture by fermentation, 113
grey moulds, 128–39

hare's ear, *89,* 90
Harposporium, 149
Helvella crispa, 89, 91
honey agaric, 37, 59
horn of plenty, *35,* 36–7
Hydnum, 49, 163–4; *H. auriscalpium, 49,* 164; *H. repandrum, 49; H. rufescens,* 49
Hypholoma fasciculare, 37–8

ink-cap, 29, 38
Isoachlya, 158
itaconic acid, manufacture by fermentation, 113

jelly jungi, 51–3
jew's ear fungus, 51

Lactarius, 36
lactic acid, manufacture by fermentation, 139
lactophenol, 165
lattice fungi, 66
Leptolegnia, 138
Leptolegniella, 158
Leveillula taurica, 103
locusts, destruction of by entomogenous fungi, 151
Lycoperdon perlatum. 63

Marasmius coniatus, 59; *M. oreades,* 39–40
Massospora, 151
Merulius lachrymans, 13, 45–8
Monacrosporium cionopagum, 147
Monilinia fructigena 11, 82
Morchella esculenta 82, *89,* 90
morel, 82, *89,* 90; morel, false, *89,* 91
Mortierella, 137
Mucor, 128–33; *M. genevensis, 132, 133; M. hiemalis, 132, 133; M. mucedo, 129, 130, 133, 137; M. plumbeus,* 129; *M. racemosus,* 137; *M. spinosus,* 129

mushroom, 11, 25–9, 38–9; mushroom spawn, 25
Mutinus caninus, 66
mycorrhiza, 54–61

nail head of apples, 96
Nectria cinnabarina, 97–8; *N. galligena*, 98
Neurospora crassa, 96; *N. sitophila*, 95–6; *N. tetrasperma*, 96
Nummularia discreta, 96

oats, loose smut of, 79
Olpidium viciae, 161
onion smut, 79
orange peel fungus, 82, 88 *89*
Otidea leporina, *89*, 90

Palaeomyces asteroxyli, 60
partial veil, 31, 32
penicillin, 11, 107–111
Penicillium, 14, 15, 105–6, *106*; *P. chrysogenum*, 11, 110; *P. notatum*, 107–10
Peziza badia, 88; *P. vesiculosa*, 88–90
Phallus impudicus, 64–6, *65*
Phoma radicis, 57, 59
Phragmidium, 77
Phycomyces, 133
Phyllactinia, 105; *P. corylea*, 103
physiologic races, of rusts, 75–6
Phytophthora infestans, 156–7
Pilobolus, 134–6, *135*, *166*; *P. kleinni*, 134; *P. longipes*, 134
pin moulds, 128–34, *130*, *132*
Piptocephalis, 139, 166
Piptoporus betulinus, 44
Podosphaera, 105
poisonous toadstools, popular methods of distinguishing, 24–25
Polyporus giganteus, 14, 45
Poria vaillanti, 48
potato blight, 13, 56–7
powdery mildews, 15, 102–5
predacious fungi, 140–9
Puccinia adoxae, 77; *P. graminis*,
13, 69–78, *71*; *P. malvacearum*, 77
puff-balls, 62–4
Pyronema omphalodes, 82–88
Pythiopsis, 159
Pythium, 13, 60

Rhizoctonia, 59
rhizomorph, 37, 46, 48
Rhizophagus, 60–1
Rhizopus, 139; *R. arrhiza*, 139; *R. oryzae*, 139; *R. stolonifer*, 128, 139
ring, 31, 32, 34
Robigus, feast of, 74–5
Russula emetica, 36
rusts, 13, 69–78

Saccharomyces carlsbergensis, 122; *S. cerevisiae*, 115–9, 124–5; *S. ellipsoideus*, 123
Saccharomycodes ludwigii, 118
Saprolegnia, 158, *159*; *S. parasitica*, 157
scale insects, 53
Schizosaccharomyces, 116, *117*; *S. octosporus*, 118
Scleroderma aurantium, 64; *S. vulgare*, 42
Septobasidium, 53
sickener, the, 36
silver leaf, 50–1
skin fungi, 50–1
smuts, 78–81, *80*
Sordaria, 106; *S. fimicola*, 94–5
Sphaerobolus, 67–8, *68*
Sphaerotheca, 105; *S. mors-uvae*, 103; *S. pannosa*, 103; *S. phyptophila*, 103
Spore prints, 30, 164
stag's horn fungus, 96, 97
Sterum hirsutum, 51; *S. purpureum*, 50–1
steroid transformations, 113–14
stinkhorn, 64–66, *65*; stinkhorn, dog's, 66
Streptomyces, 112; *S. aureofaciens*, 112; *S. griseus*, 112

streptomycin, 112
Stylopage, 144; *S. cymosa*, *142*; *S. grandis*, 142, *143*; *S. hadra*, 142
sulphur tuft, 37–8
Syncephalastrum, 139
Syncephalis, 139

Taphrina, 88
Thamnidium, 137, 138
Tilletia, 78; *T. caries*, 79; *T. foetida*, 79
tooth fungi, 49–50, *49*
Tremella mesenterica, 53
truffle, 64, 91–2; truffle, false, 42
Tuber aestivum, 92; *T. rufrum*, 92

Uncinula, 105; *U. necator*, 103
universal veil, *31*, 32
Urocystis cepulae, 79
Uromyces, 77

Urophylyctis alfalfae, 160
Ustilago avenae, 79; *U. maydis*, 79

Verticillium, *19*
vine mildew, 153–156, *155*
volva, *31*, 32, 34

wart disease of potatoes, 160
water moulds, 157–60
wheat, stinking smut of, 79
white moulds, 128–39
wine making, 122–4
wood mushroom, 38–9

Xerotus javanicus, 59
Xylosphaera hypoxlon, 96, 97

yeast, 115–27
Zoopage, 144
Zygorrhinchus, 133–4